The Complete Guide to Mold Making with SOLIDWORKS 2022

By Paul Tran, Sr. Certified SOLIDWORKS Instructor

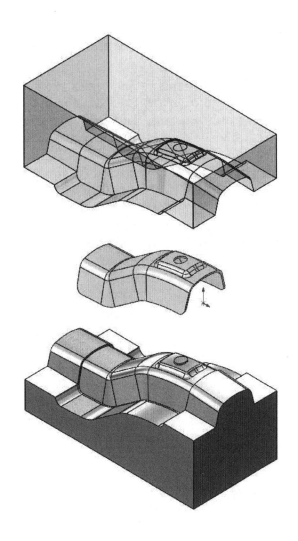

SDC
PUBLICATIONS

SDC Publications
P.O. Box 1334
Mission, KS 66222
913-262-2664
www.SDCpublications.com
Publisher: Stephen Schroff

ISBN-13: 978-1-63057-483-3
ISBN-10: 1-63057-483-X

Printed and bound in the United States of America.

Acknowledgments

Thanks as always to my wife Vivian and my daughter Lani for always being there and providing support and honest feedback on all the chapters in the textbook.

I would like to give a special thanks to Karla Werner for her editing and corrections. Additionally, thanks to Kevin Douglas and Peter Douglas for writing the forewords.

I also have to thank SDC Publications and the staff for its continuing encouragement and support for this edition of **The Complete Guide to Mold Making with SOLIDWORKS 2022**. Thanks also to Tyler Bryant for putting together such a beautiful cover design.

Finally, I would like to thank you, our readers, for your continued support. It is with your consistent feedback that we were able to create the lessons and exercises in this book with more detailed and useful information.

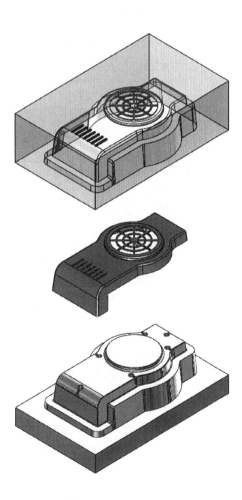

Foreword

For more than two decades, I have been fortunate to have worked in the fast-paced, highly dynamic world of mechanical product development providing computer-aided design and manufacturing solutions to thousands of designers, engineers, and manufacturing experts in the western US. The organization where I began this career was US CAD in Orange County, CA, one of the most successful SOLIDWORKS Resellers in the world. My first several years were spent in the sales organization prior to moving into middle management and ultimately President of the firm. In the mid-1990s is when I met Paul Tran, a young, enthusiastic instructor who had just joined our team.

Paul began teaching SOLIDWORKS to engineers and designers of medical devices, automotive and aerospace products, high tech electronics, consumer goods, complex machinery and more. After a few months of watching him teach and interacting with students during and after class, it was becoming pretty clear – Paul not only loved to teach, but his students were the most excited with their learning experience than I could ever recall from previous years in the business. As the years began to pass and thousands of students had cycled through Paul's courses, what was eye opening was Paul's continued passion to educate as if it were his first class and students in every class, without exception, loved the course.

Great teachers not only love their subject, but they love to share that joy with students – this is what separates Paul from others in the world of SOLIDWORKS Instruction. He always has gone well beyond learning the picks & clicks of using the software, to best practice approaches to creating intelligent, innovative, and efficient designs that are easily grasped by his students. This effective approach to teaching SOLIDWORKS has translated directly into Paul's many published books on the subject. His latest effort with SOLIDWORKS 2020-2021-2022 is no different. Students that apply the practical lessons from basics to advanced concepts will not only learn how to apply SOLIDWORKS to real world design challenges more quickly but will gain a competitive edge over others that have followed more traditional approaches to learning this type of technology.

As the pressure continues to rise on U.S. workers and their organizations to remain competitive in the global economy, raising not only education levels but technical skills is paramount to a successful professional career and business. Investing in a learning process towards the mastery of SOLIDWORKS through the tutelage of the most accomplished and decorated educator and author in Paul Tran will provide a crucial competitive edge in this dynamic market space.

Kevin Douglas
Vice President Sales/Board of Advisors, GoEngineer

Preface

I first met Paul Tran when I was busy creating another challenge in my life. I needed to take a vision from one man's mind, understand what the vision looked like, how it was going to work and comprehend the scale of his idea. My challenge was I was missing one very important ingredient, a tool that would create a picture with all the moving parts.

Research led me to discover a great tool, SOLIDWORKS. It claimed to allow one to make 3D components, in picture quality, on a computer, add in all moving parts, assemble it, and make it run, all before money was spent on bending steel and buying parts that may not fit together. I needed to design and build a product with thousands of parts, make them all fit and work in harmony with tight tolerances. The possible cost implications of failed experimentation were daunting.

To my good fortune, one company's marketing strategy of selling a product without an instruction manual and requiring one to attend an instructional class to get it, led me to meet a communicator who made it all seem so simple.

Paul Tran has worked with and taught SOLIDWORKS as his profession for more than 35 years. Paul knows the SOLIDWORKS product and manipulates it like a fine musical instrument. I watched Paul explain the unexplainable to baffled students with great skill and clarity. He taught me how to navigate the intricacies of the product so that I could use it as a communication tool with skilled engineers. *He teaches the teachers*.

I hired Paul as a design engineering consultant to create the thousands of parts for my company's product. Paul Tran's knowledge and teaching skill has added immeasurable value to my company. When I read through the pages of these manuals, I now have an "instant replay" of his communication skill with the clarity of having him looking over my shoulder - *continuously*. We can now design, prove and build our product and know it will always work and not fail. Most important of all, Paul Tran helped me turn a blind man's vision into reality and a monument to his dream.

Thanks Paul.

These books will make dreams come true and help visionaries change the world.

Peter J. Douglas
CEO, Cake Energy, LLC

Images courtesy of C.A.K.E. Energy Corp., designed by Paul Tran

About the Author

The Complete Guide to Mold Making with SOLIDWORKS 2022 is comprised of lessons based on the feedback from Paul's former tool-die maker students and engineering professionals. Paul has more than 35 years of experience in the fields of mechanical and manufacturing engineering; 2/3 of those years were spent in teaching and supporting the SOLIDWORKS software and its add-ins. As an active Sr. SOLIDWORKS instructor and design engineer, Paul has worked and consulted with hundreds of reputable companies including IBM, Intel, NASA, US-Navy, Boeing, Disneyland, Medtronic, Edwards Lifesciences, Microvention, Oakley, Kingston, community colleges, universities, and many others. Today, he has trained more than 13,000 engineering professionals, and given guidance to half of the number of Certified SOLIDWORKS Professionals and Certified SOLIDWORKS Expert (CSWP & CSWE) in the state of California.

Every lesson in this book was created based on the actual products. Each of these projects have been broken down and developed into easy and comprehendible steps for the reader. Furthermore, every mold design is explained very clearly in short chapters, ranging from 15 to 30 pages. Each and every single step comes with the exact screen shot to help you understand the main concept of each design more easily. Learn the Mold Designs at your own pace, as you progress from simple core and cavity creation and then to more complex mold design challenges.

About the Training Files

The files for this textbook are available for download on the publisher's website at www.SDCpublications.com/downloads/978-1-63057-483-3. They are organized by the chapter numbers and the file names that are normally mentioned at the beginning of each chapter or exercise. In the **Completed Parts** folder you will also find copies of the parts, assemblies and drawings that were created for cross referencing or reviewing purposes.

It would be best to make a copy of the content to your local hard drive and work from these documents; you can always go back to the original training files location at anytime in the future, if needed.

Who this book is for

This book is for the beginner-level to advanced user, who is already familiar with the SOLIDWORKS program. To get the most out of this mold designs book it is strongly recommended that you have studied and completed all the lessons in the Basic and Advance textbooks. It is also a great resource for the more CAD literate individuals who want to expand their knowledge of the different features that SOLIDWORKS 2022 has to offer.

The organization of the book

The chapters in this book are organized in the logical order in which you would learn the Mold Designs using SOLIDWORKS 2022. Each chapter will guide you through some different tasks, from designing or repairing a mold, to developing complex parting lines; from making a core in the part mode to advancing through more complex tasks in the assembly mode.

You will also learn to work with SOLIDWORKS Plastics to simulate how melted plastics flows during the injection molding process and analyze the thick-thin wall regions to predict defects on plastic parts and molds.

The conventions in this book

This book uses the following conventions to describe the actions you perform when using the keyboard and mouse to work in SOLIDWORKS 2022:

Click: means to press and release the mouse button. A click of a mouse button is used to select a command or an item on the screen.

Double Click: means to quickly press and release the left mouse button twice. A double mouse click is used to open a program or show the dimensions of a feature.

Right Click: means to press and release the right mouse button. A right mouse click is used to display a list of commands, a list of shortcuts that is related to the selected item.

Click and Drag: means to position the mouse cursor over an item on the screen and then press and hold down the left mouse button; still holding down the left button, move the mouse to the new destination and release the mouse button. Drag and drop makes it easy to move things around within a SOLIDWORKS document.

Bolded words: indicates the action items that you need to perform.

Italic words: Side notes and tips that give you additional information, or to explain special conditions that may occur during the course of the task.

Numbered Steps: indicates that you should follow these steps in order to successfully perform the task.

Icons: indicates the buttons or commands that you need to press.

SOLIDWORKS 2022

SOLIDWORKS 2022 is program suite, or a collection of engineering programs that can help you design better products faster. SOLIDWORKS 2022 contains different combinations of programs; some of the programs used in this book may not be available in your suites.

Start and exit SOLIDWORKS

SOLIDWORKS allows you to start its program in several ways. You can either double click on its shortcut icon on the desktop or go to the Start menu and select the following: All Programs / SOLIDWORKS 2022 / SOLIDWORKS or drag a SOLIDWORKS document and drop it on the SOLIDWORKS shortcut icon.

Before exiting SOLIDWORKS, be sure to save any open documents, and then click File / Exit; you can also click the X button on the top right of your screen to exit the program.

Using the Toolbars

You can use toolbars to select commands in SOLIDWORKS rather than using the drop-down menus. Using the toolbars is normally faster. The toolbars come with commonly used commands in SOLIDWORKS, but they can be customized to help you work more efficiently.

To access the toolbars, either right click in an empty spot on the top right of your screen or select View / Toolbars.

To customize the toolbars, select Tools / Customize. When the dialog pops up, click on the Commands tab, select a Category, then drag an icon out of the dialog box and drop it on a toolbar that you want to customize. To remove an icon from a toolbar, drag an icon out of the toolbar and drop it into the dialog box.

Using the task pane

The task pane is normally kept on the right side of your screen. It displays various options like SOLIDWORKS resources, Design library, File explorer, Search, View palette, Appearances and Scenes, Custom properties, Built-in libraries, Technical alerts, and news, etc.

The task pane provides quick access to any of the mentioned items by offering the drag and drop function to all of its contents. You can see a large preview of a SOLIDWORKS document before opening it. New documents can be saved in the task pane at anytime, and existing documents can also be edited and re-saved. The task pane can be resized, closed, or moved to different locations on your screen if needed.

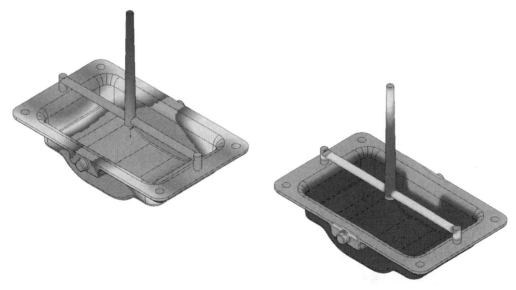

It is recommended that the user should have some mechanical or tooling design experience, completion of the Essentials training course or have at least 6 months of using SOLIDWORKS

TABLE OF CONTENTS

The Complete Guide to Mold Making with SOLIDWORKS 2022

Chapter 7: Manual Parting Lines   **7-1**

Chapter 12: Plastics Flow Analysis   12-1

Glossary

Index by chapters

SOLIDWORKS 2022 Quick-Guides:

Quick Reference Guide to SOLIDWORKS 2022 Command Icons and Toolbars.

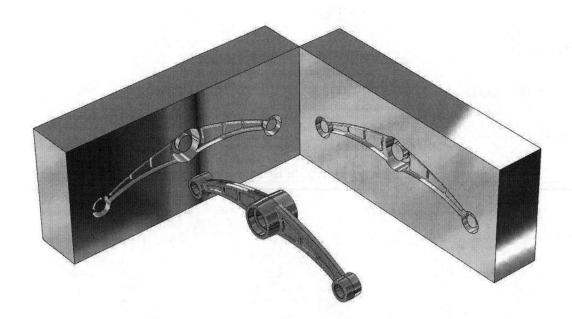

CHAPTER 1

Plastic Part Design 1

When designing a plastic part there
are a few key points to keep in mind:
First, the wall thickness should be uniform throughout the part and have adequate drafts.
Avoid solid features as thicker areas will shrink and warp more than thin areas.
Second, all corners, both internal and external, should be filleted to reduce sink marks.
Sharp corners can crack, and they require a longer cooling time.
Lastly, add ribs to increase rigidity of the plastic part. Other options such as parting lines,
shut-off surfaces, etc., will be discussed in other chapters.

1. Starting a new part document:

Select **File, New**.

Select the **Part** template and click **Open**.

Set the Drafting Standards to **ANSI** and the
number of decimals to **3 Places**.

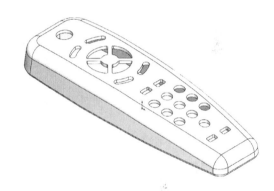

2. Creating the parent sketch:

The main sketch, also called the parent sketch, defines the overall shape and size
of the part. It also dictates how the model is oriented with respect to the Front, Top,
and Right (X,Y,Z) directions.

Select the Top plane
and open a **new sketch**.

Sketch the profile
shown using
the **Line**, the
3-Point-Arc and
the **Mirror**
commands.

Add dimensions
to fully define
the sketch.

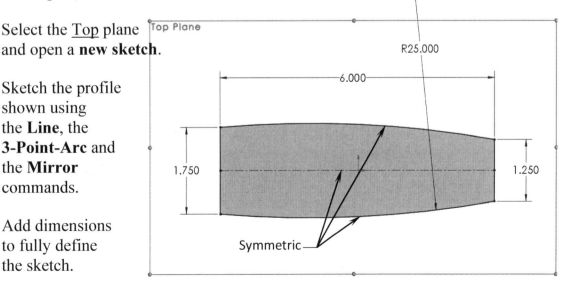

3. Extruding the main body:

Switch to the **Features** tab and click **Extruded Boss-Base**.

For Direction 1, use the default **Blind** type.

For Extrude Depth, enter: **.750in**.

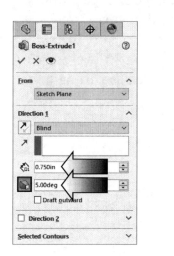

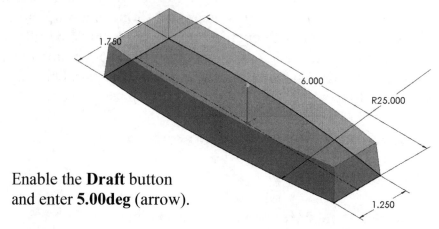

Enable the **Draft** button
and enter **5.00deg** (arrow).

Click **OK**.

4. Making the upper curved cut:

Select the Front plane and open a **new sketch**.

Sketch a **3-Point-Arc** and add the **Coincident** relations to both ends as noted.

Add the dimensions shown to fully define the sketch.

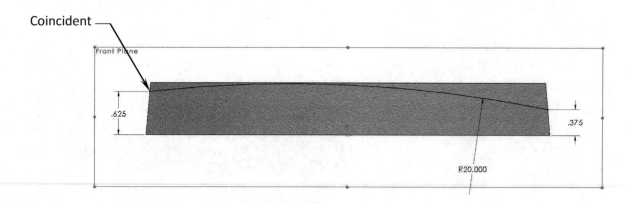

Switch to the **Features** tab.

Click **Extruded Cut**.

For Direction 1, select **Through All Both** (arrow).

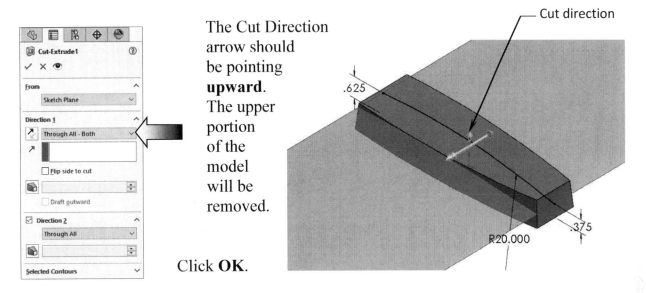

The Cut Direction arrow should be pointing **upward**. The upper portion of the model will be removed.

Click **OK**.

5. Adding the .500in fillets:

Click **Fillet** on the Features tab.

For Fillet Type, use the default **Constant Size Radius** option.

For Radius size, enter **.500in** (arrow).

For Items to Fillet, select the **2 edges** as indicated.

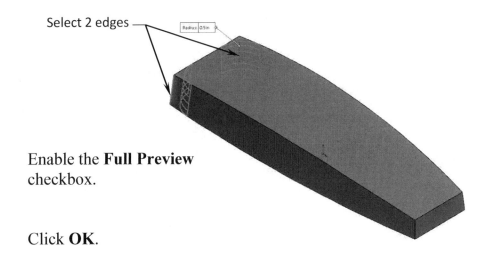

Enable the **Full Preview** checkbox.

Click **OK**.

6. Adding the .250in fillets:

Click **Fillet** again.

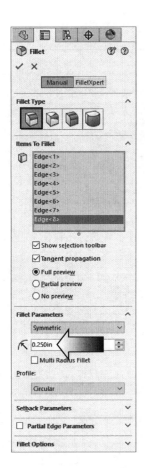

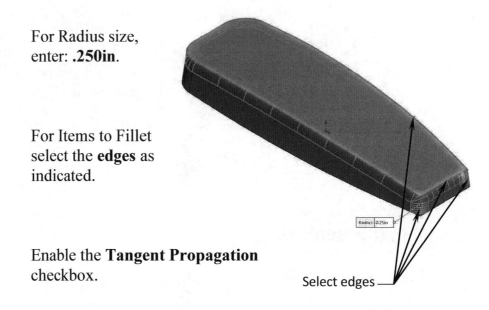

For Fillet Type, use the default **Constant Size Radius** option.

For Radius size, enter: **.250in**.

For Items to Fillet select the **edges** as indicated.

Enable the **Tangent Propagation** checkbox.

Select edges

Click **OK**.

7. Shelling the part:

Click the **Shell** command on the Features tab.

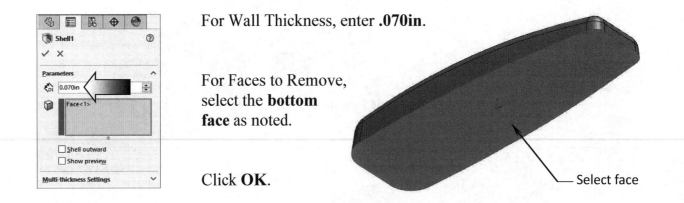

For Wall Thickness, enter **.070in**.

For Faces to Remove, select the **bottom face** as noted.

Click **OK**.

Select face

8. Making the recess sketch:

Select the <u>Top</u> plane and open a **new sketch**.

The recess
feature can
help align
the two
halves and
also make
the part
appear
more attractive.

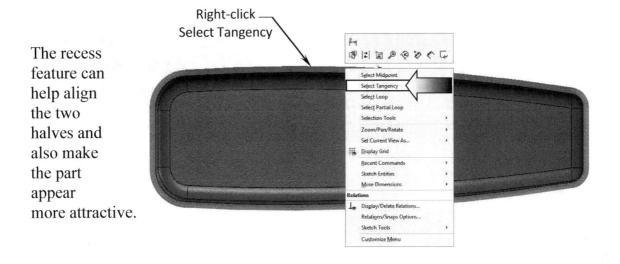

Right-click on the <u>outer edge</u> and
pick **Select Tangency** (arrow).

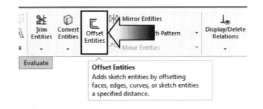

Click **Offset Entities** (arrow).

For Offset Distance, enter **.035in** (arrow).

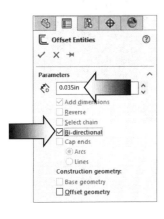

Enable the **Bi-Directional** checkbox. The preview graphics
show 2 profiles are being created; each one is .035" apart or
.070" total distance.

Click **OK**.

9. Making the recess cut:

Switch to the **Features** tab.

Click **Extruded Cut**.

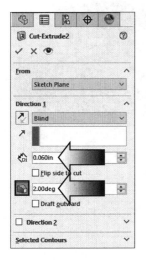

For Direction 1, use the default **Blind** type.

For Extrude Depth, enter **.060in**. (arrow).

Enable the **Draft** On/Off button and enter **2.00deg** (arrow).

Use the default <u>inward draft</u> direction (see detail view below).

Click **OK**.

The recess cut is created and it has a
2 degrees draft around its perimeter.

The sketch of the buttons will be used to create the holes in the next few steps.

10. Copying a sketch:

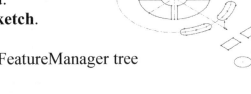

Open a part document named:
Remote-Control Buttons Sketch.

Select the **Sketch1** from the FeatureManager tree
and press **Control+C**.

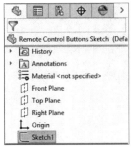

Press **Control+Tab** to switch back to the previous
document.

Select the **Top** plane and
press **Control+V**.

Edit the sketch of the buttons and add a
Coincident between the end point of the
centerline and the origin to fully define the sketch.

Change to the **Features** tab and select **Extruded Cut**.

Select the **Offset** option under the **Extrude From** drop down.

For Offset Distance, enter **.750in**.

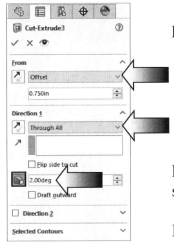

For Direction 1,
select **Through All**.

Enable the **Draft** button
and enter **2.00deg**.

Click **OK**.

11. Creating the mounting bosses:

Select the <u>face</u> as indicated and open a **new sketch**.

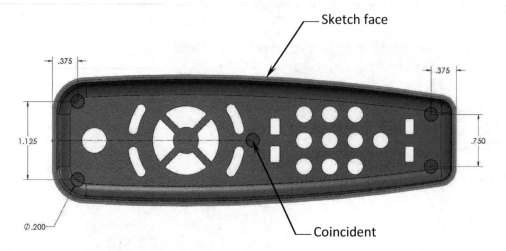

Sketch **5 Circles** and add the dimensions shown above.

The center of the circle in the middle is coincident to the origin.

Add an **Equal** relation to the 5 circles to fully define the sketch.

Switch to the **Features** tab and select **Extruded Boss-Base**.

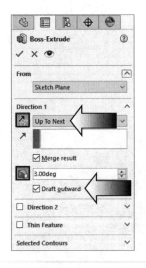

For Direction 1, click **Reverse** and select the **Up-To-Next** option from the drop-down list.

Enable the **Merge Result** checkbox.

Click the **Draft On/Off** button and enter **3.00deg**.

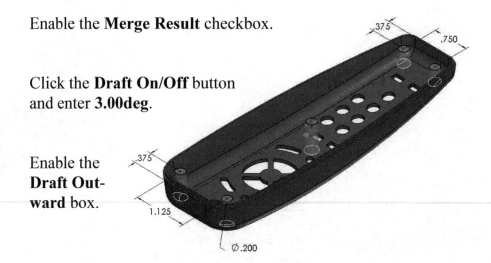

Enable the **Draft Outward** box.

Click **OK**.

12. Adding the support ribs:

Select the <u>top face</u> of one of the mounting bosses and open a **new sketch**.

Sketch the **Rectangles** shown in the image below. Use the **Mirror** option where applicable to keep them symmetric to each other.

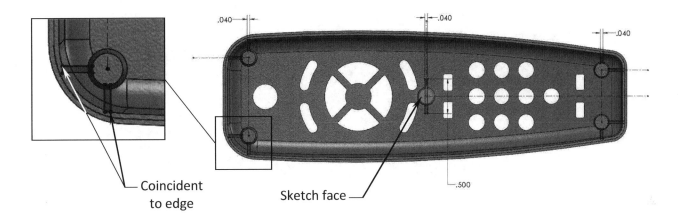

Coincident
to edge

Sketch face

Sketch the additional **Centerlines** and add **Symmetric** relations to the rectangles.

Add the dimensions and relations needed to fully define the sketch.

Switch to the **Features** tab and click **Extruded Boss-Base**.

For Direction 1, click **Reverse** and select **Up-To-Next**.

Click the **Draft** button and enter **2.00deg** for draft.

Enable the Draft **Outward** box.

Click **OK**.

13. Adding the mounting holes:

Select the <u>top face</u> of one of the mounting bosses and open a **new sketch**.

Sketch **5 circles** centered on the circular edges of the mounting bosses.

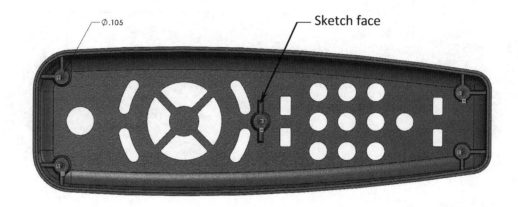

Ø.105 — Sketch face

Add an **Equal** relation to the 5 circles. Ensure that the centers of the circles are **Coincident** to the centers of the circular bosses. (Concentric relations can also be used between the circles and the circular edges of the bosses.)

Switch to the **Features** tab and click **Extruded Cut**.

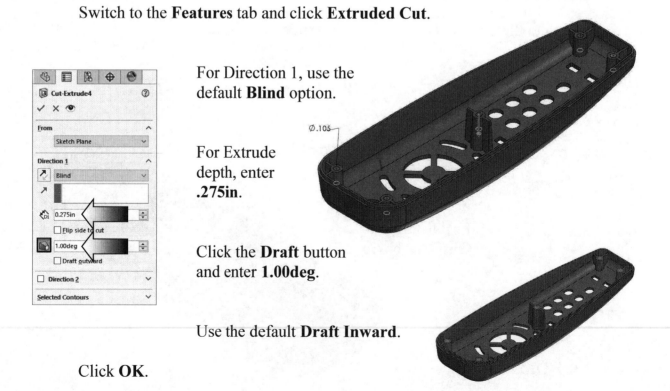

For Direction 1, use the default **Blind** option.

For Extrude depth, enter **.275in**.

Click the **Draft** button and enter **1.00deg**.

Use the default **Draft Inward**.

Click **OK**.

14. Adding the rib fillets:

Select the **Fillet** command on the Features tab.

For Fillet Type, use the default **Constant Size Radius** option.

For Radius Size, enter **.020in**. (arrow).

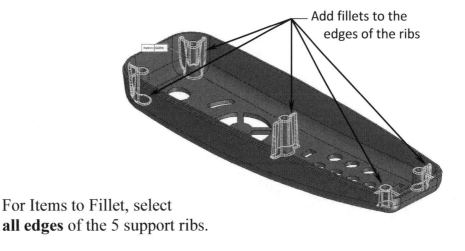

Add fillets to the edges of the ribs

For Items to Fillet, select **all edges** of the 5 support ribs.

Create 2 different fillets if needed (some edges may not blend correctly with the adjacent edges).

Click **OK**.

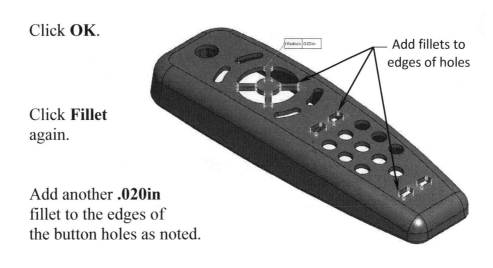

Add fillets to edges of holes

Click **Fillet** again.

Add another **.020in** fillet to the edges of the button holes as noted.

Click **OK**.

(The .020in fillets can be combined into one fillet if preferred.)

15. Assigning material:

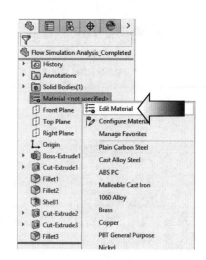

Right-click **Material** and select **Edit Material**.

Expand the **Plastics** folder and select **ABS** (arrow).

The density is: **0.037 pounds per cubic inch**.

Click **Apply** and **Close**.

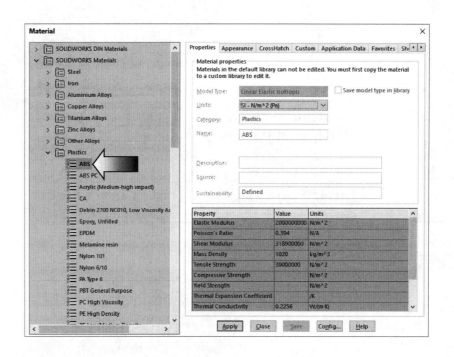

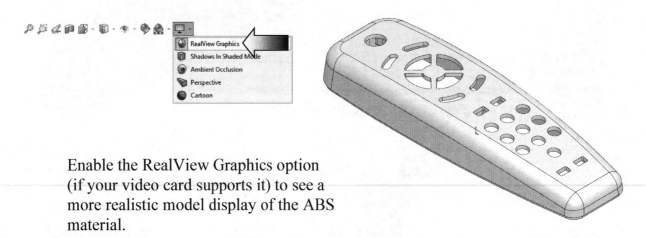

Enable the RealView Graphics option
(if your video card supports it) to see a
more realistic model display of the ABS
material.

16. Changing the part color:

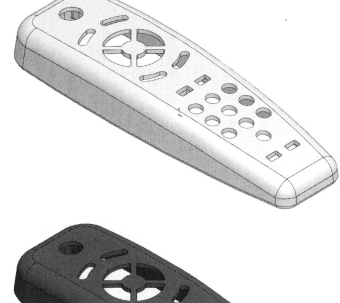

The default color for SOLIDWORKS ABS is white.
We will change it to gray for this lesson.

Click **Edit Appearance** on the
View Heads-Up toolbar (or
select: **Edit, Appearance,
Appearance**).

Under the **Color/Image** tab,
select the **Gray** color (arrow).

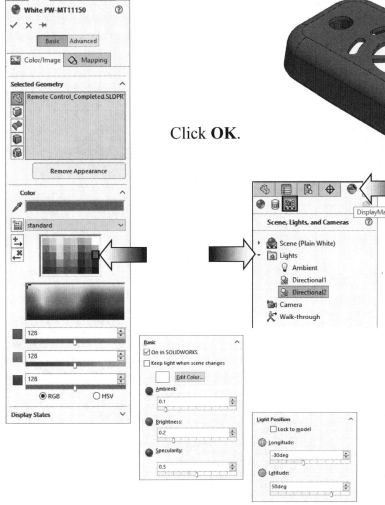

Click **OK**.

If default scene is too dark, you
can adjust it by switching to the
DisplayManager tab, expanding
the Lights folder, double-clicking
on one of the lights and adjusting
the Ambient or the Brightness
lights to your preference.

17. Calculating the mass:

Switch to the **Evaluate** tab.

Click **Mass Properties** (arrow).

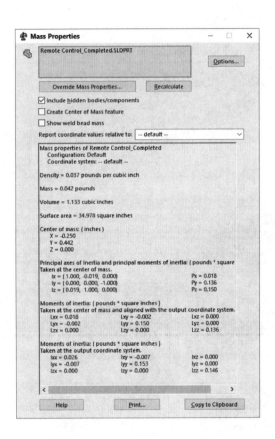

A dialog box appears listing the following mass properties:

* Density
* Mass
* Volume
* Surface area
* Center of mass
* Principal axes of inertia
* Moments of inertia and products of inertia

In the graphics area, a single-colored triad indicates the principal axis and center of mass of the model.

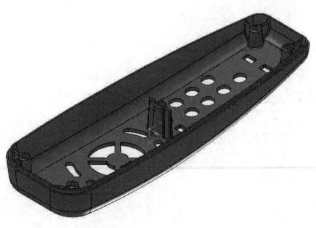

*Optional: Click **Options** and make any changes in the Mass/Section Property Options dialog box such as pounds, Inches, and click OK. The results update accordingly.*

18. Applying the scale factor:

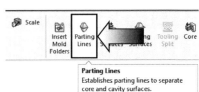

The model is completed. We will now
design the Core & Cavity Mold for this part.
The first step is to scale the part to accommodate the shrinkage of the material.

Change to the **Mold Tools** tab and click **Scale** (arrow).

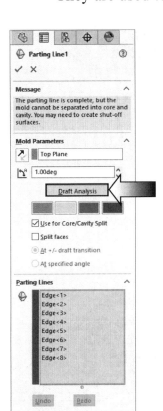

For Scale About, select the **Centroid** option.

Enable the **Uniform Scaling** checkbox.

For Scale Factor, enter **1.02** (2% larger).

Click **OK**.

19. Creating the parting lines:

The parting lines run around the edges of the
plastic part, between the core and the cavity surfaces.
They are used to create the parting surfaces and to separate the surfaces.

Click **Parting Lines** (arrow) on the Mold Tools tab.

For Direction of Pull, select the **Top** plane from the Feature tree.

For Draft Angle, enter **1.00deg** and click **Draft Analysis** (arrow).

The bottom edges of the model are selected automatically. They
will be used as the parting lines for this part.

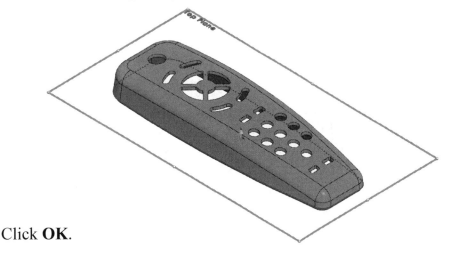

Click **OK**.

20. Creating the shut-off surfaces:

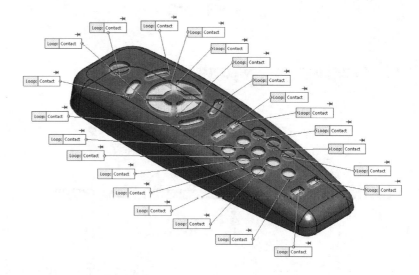

To split a tooling block into a pair of core and cavity blocks, two surfaces need to be completed (a core surface and a cavity surface) without any through holes.
The shut-off surfaces close up all through holes by creating a surface patch along the edges that form a continuous loop.

Click **Shut-Off Surfaces** (arrow) on the Mold Tools tab.

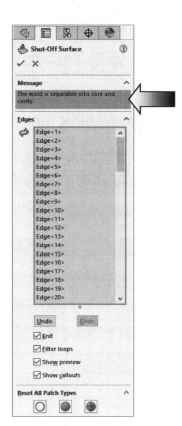

The edges of all through holes are selected automatically.

A green colored message appears on the upper left corner indicating the mold is separable into core and cavity.

Enable the options:

       * **Knit**
       * **Filter Loops**
       * **Show Preview**
       * **Show Callout**

Click **OK**.

21. Creating the parting surfaces:

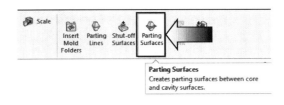

The parting surfaces, which is extended from the parting lines, are used to separate the mold cavity from the core.

Click **Parting Surfaces** (arrow).

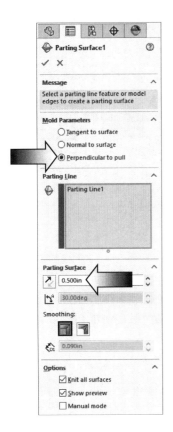

For Mold Parameters, select **Perpendicular to Pull** (arrow).

The **Parting Line1** is selected automatically.

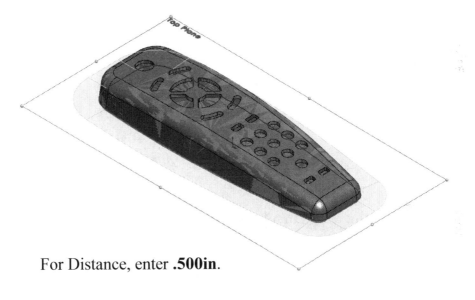

For Distance, enter **.500in**.

For Smoothing, use the default **Sharp** option.

Enable the options:

      * **Knit All Surfaces**
      * **Show Preview**

Click **OK**.

A parting surface is created by protruding outward from the parting lines.

22. Adding a reference plane:

The reference plane is used to create the sketch of the mold block and also set the height of the interlock surface during the tooling split.

Click **Plane** (or select **Insert, Reference Geometry, Plane**).

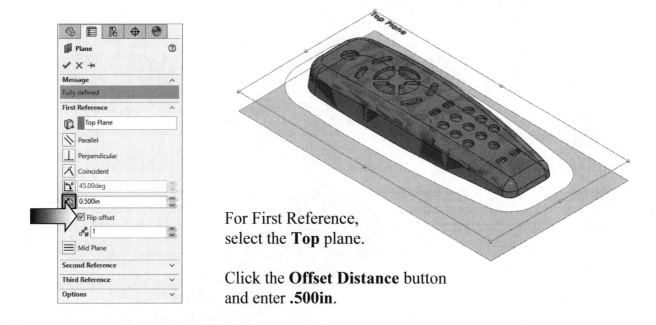

For First Reference,
select the **Top** plane.

Click the **Offset Distance** button
and enter **.500in**.

Click the **Flip Offset** checkbox to place the new plane <u>below</u> the Top plane.

Click **OK**.

Open a **new sketch**
on the <u>Plane1</u>.

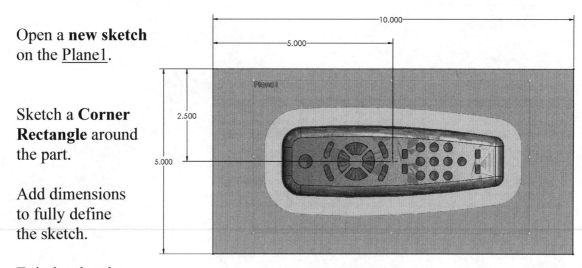

Sketch a **Corner Rectangle** around the part.

Add dimensions
to fully define
the sketch.

<u>Exit</u> the sketch.

23. Inserting a tooling split:

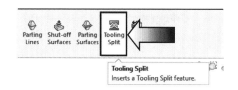

The tooling split command uses a sketch to create the mold block and splits it into core and cavity blocks.

To help prevent the core and cavity blocks from shifting, an interlock surface is added along the perimeter of parting surfaces.

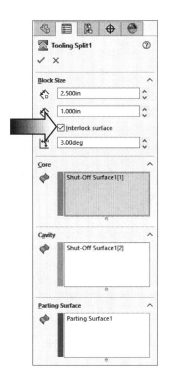

Select the sketch of the **rectangle** and click **Tooling Split**.

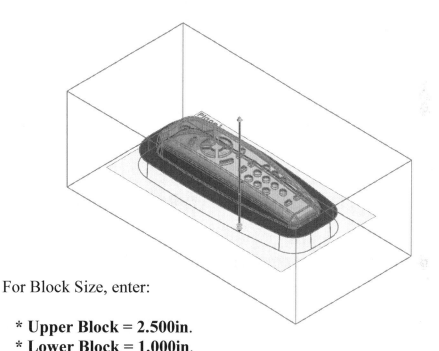

For Block Size, enter:

    * **Upper Block = 2.500in**.
    * **Lower Block = 1.000in**.

Enable the **Interlock Surface** checkbox.

Enter **3.00deg** for Draft Angle.

Click **OK**.

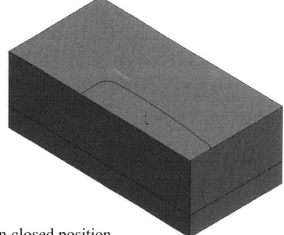

The Core and Cavity Blocks are shown in closed position.

24. **Separating the mold blocks:**

There are a few options available to separate the mold blocks, but we will use the Move/Copy Bodies command for this lesson. The moves are created as features and stored on the FeatureManager; they can be suppressed to collapse the mold blocks quickly and easily.

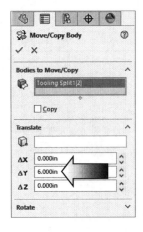

Click **Move/Copy Bodies** (or select: **Insert, Features, Move/Copy Bodies**).

For Bodies to Move/Copy, select the **upper block** (the cavity).

Click in the <u>Delta Y</u> field and enter **6.000in**. for distance.

Click **OK**.

The upper block moves upward 6 inches along the Y direction.

Click **Move/Copy Bodies** again.

For Bodies to Move/Copy, select the **lower block** (the core).

Click in the <u>Delta Y</u> field and enter **-4.500in**. for distance.

Click **OK**.

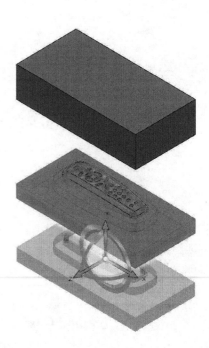

The lower block moves downward 4.5 inches along the -Y direction.

25. Hiding the references:

The references such as parting lines, shut-off surfaces, parting surfaces are put away at this point for clarity.

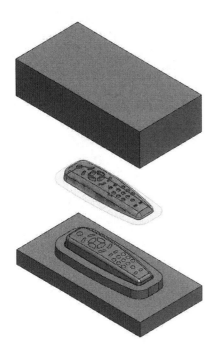

To hide the parting line, select: **View, Hide/Show, Parting Lines** (arrow).

To hide the surfaces, click the Surface Bodies folder and select: **Hide** (arrow).

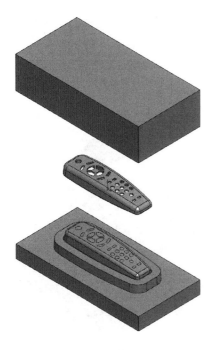

Rename the solid bodies to:

  * **Plastic part**

  * **Cavity Block**

  * **Core Block**

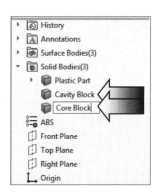

26. Assigning material to the mold blocks:

The Plastic Part already has the ABS material assigned to it previously.

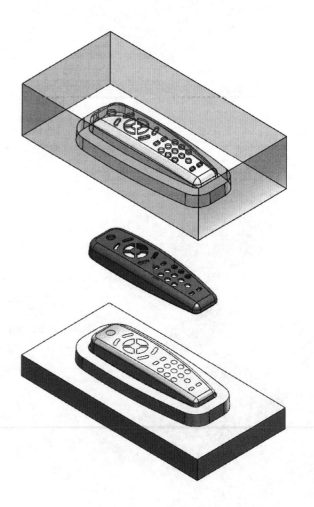

Right-click the Cavity Block, select **Material, Plain Carbon Steel**.

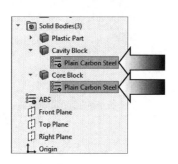

Right-click the Core Block, select **Material, Plain Carbon Steel** (arrow).

Suppress the 2 move features to collapse the mold blocks.

Right-click the Cavity Block and select: **Change Transparent**.

Also make the Core Block transparent.

Un-suppress the 2 move features to explode the 2 mold blocks.

27. Saving your work:

Select **File, Save As**.

Enter **Plastic Part Design_Completed** for the file name.

Click **Save**.

Close all documents.

Exercise – Plastic Part Design 2

When designing plastic parts for injection molding, applying draft to the faces of the part is important to improve the moldability of your part. Without draft, parts run the risk of poor cosmetic finishes, and may warp or break, due to molding stresses caused by the plastic cooling. If draft is missing, the part may not be able to be ejected from the mold, and can also damage the mold as well.

1. Opening a part document:

Select **File, Open**.

Select the part document named **Plastic Part Design 2.sldprt** and click **Open**.

This part document has been partially modeled. We will add the finishing touches and create a set of core and cavity mold for this part towards the end.

2. Extruding the Fins:

Select the **Sketch17** from the FeatureManager tree and click **Extruded Boss-Base**.

Click the **Reverse Direction** button.

For Direction 1, select **Up-To-Next** from the list.

Enable the **Merge Result** checkbox.

Enable the **Draft On-Off** button.

Enter **1.00°** click **Draft Outward**.

Click **OK.**

3. Adding chamfers:

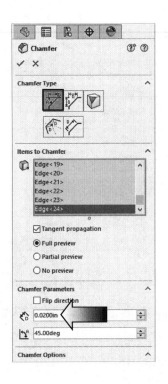

Select **Chamfer** from the Fillet drop-down list.

Use the default **Angle and Distance** option.

For Items to Chamfer, select **24 edges** on both sides, as noted.

For Chamfer Depth, enter **.020in**.

For Chamfer Angle, enter **45°**.

Click **OK**.

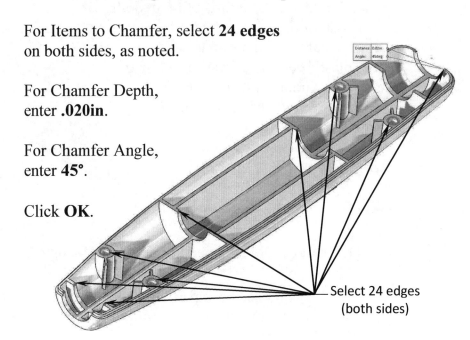

Select 24 edges
(both sides)

4. Adding fillets:

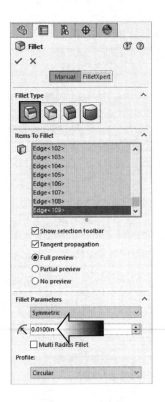

Click **Fillet** and use the default **Constant Size Radius** option.

For Items to Fillet, select all edges of the ribs and mounting bosses, except for some of the edges as indicated.

For Radius, enter **.010in**.

Click **OK**.

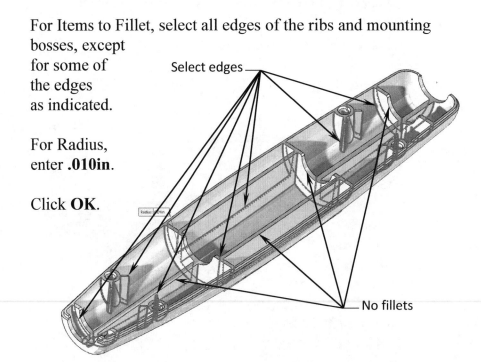

Select edges

No fillets

5. Applying scale:

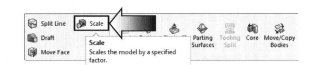

Switch to the **Mold Tools** tab.

Click **Scale**.

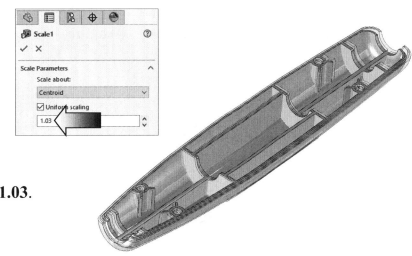

For Scale About, use the default **Centroid** option.

Enable the **Uniform-Scaling** checkbox.

For Scale Factor, enter **1.03**.

Click **OK**.

6. Creating the parting line:

Select the **Parting Lines** command from the **Mold Tools** tab.

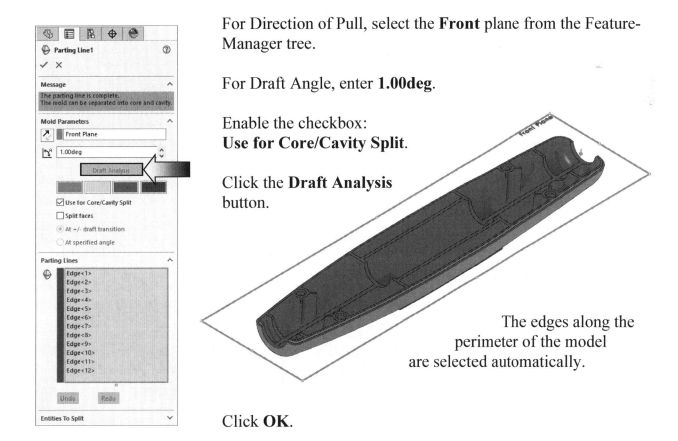

For Direction of Pull, select the **Front** plane from the Feature-Manager tree.

For Draft Angle, enter **1.00deg**.

Enable the checkbox: **Use for Core/Cavity Split**.

Click the **Draft Analysis** button.

The edges along the perimeter of the model are selected automatically.

Click **OK**.

7. Adding the parting surfaces:

Since there is no through hole in the model, we can skip the step to shut-off the surfaces and move on to creating the parting surfaces.

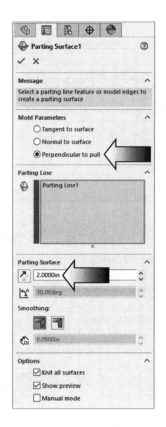

Click **Parting Surface**.

For Mold Parameters, select the option **Perpendicular to Pull**.

The **Parting Line** is selected automatically.

For Parting Surface Distance enter **2.00in**.

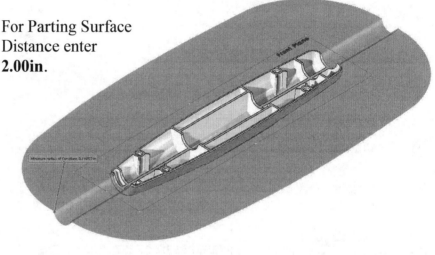

Click **OK**.

8. Sketching the mold profile:

Open a <u>new sketch</u> on the **Front** plane.

Select all <u>outer edges</u> of the Parting Surface1 and press **Convert-Entities**.

Additionally, sketch a construction rectangle and add the dimensions shown.

The rectangle will be used to trim the mold blocks in the next step.

Convert all outer edges

9.0000

4.0000

1.1250

Exit the sketch.

9. Creating a tooling split:

Click **Tooling Split** and select the sketch that was created in the last step.

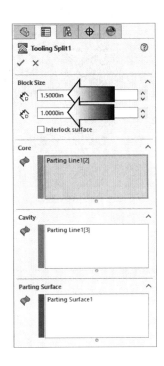

For Block Size, enter:
1.50in for upper block
1.00in for lower block

The Core,
Cavity,
and Parting
Surface
are selected
automatically.

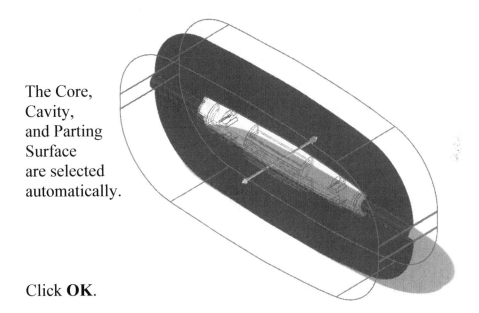

Click **OK**.

10. Trimming the mold blocks:

In some cases it might be easier to trim the mold blocks to their final shape and size, after the fact.

Show the **Sketch18** under the Tooling Split feature.

Open a **new sketch** on the face as indicated.

Select the rectangle and press **Convert Entities**.

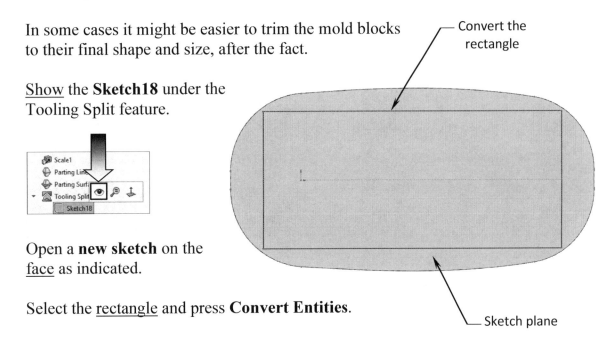

Convert the rectangle

Sketch plane

Switch to the **Features** tab and click **Extruded Cut**.

For Direction 1 select **Through All – Both**.

Enable the check box: **Flip Side To Cut**.

Under Feature-Scope, select the **2 mold blocks**.

Click **OK**.

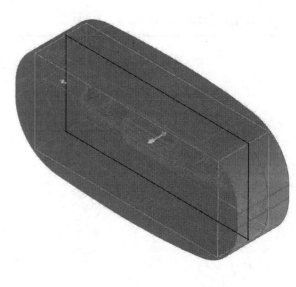

11. Moving the front mold block:

Click **Move/Copy Bodies** or select: **Insert, Features, Move/Copy Bodies**.

For Bodies to Move/Copy, select the **Cavity Block**.

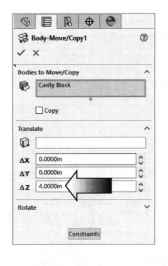

Drag the **Z Arrow** outward, about **4.00in**.

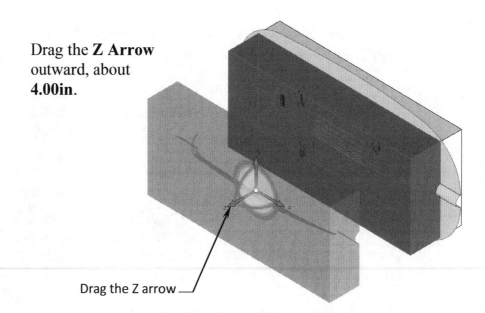

Drag the Z arrow

Click **OK**.

12. Moving the rear mold block:

Select the **Move/Copy Bodies** command once again.

Select the **Core Block** and drag the **Z Arrow** backwards approximately **-3.75in**.

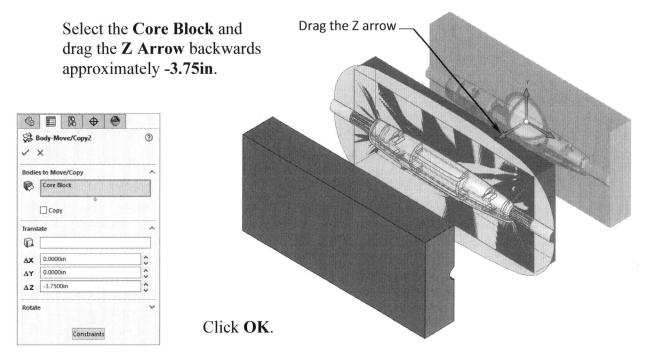

Drag the Z arrow

Click **OK**.

13. Hiding the reference surfaces:

The reference surfaces can now be put away.

Click the Surface Bodies folder and select **Hide**.

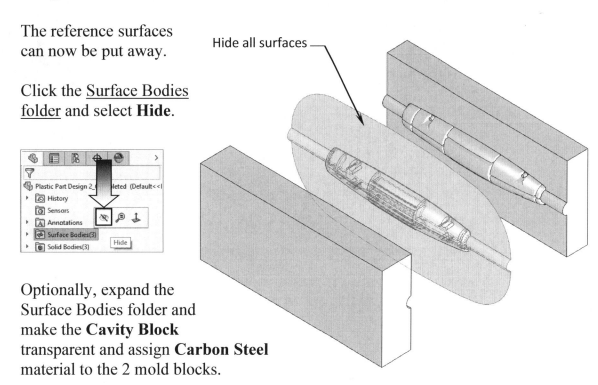

Hide all surfaces

Optionally, expand the Surface Bodies folder and make the **Cavity Block** transparent and assign **Carbon Steel** material to the 2 mold blocks.

14. Saving your work:

Select **File, Save As**.

For file name, enter: **Plastic Part Design 2_Completed.sldprt**

Click **Save**.

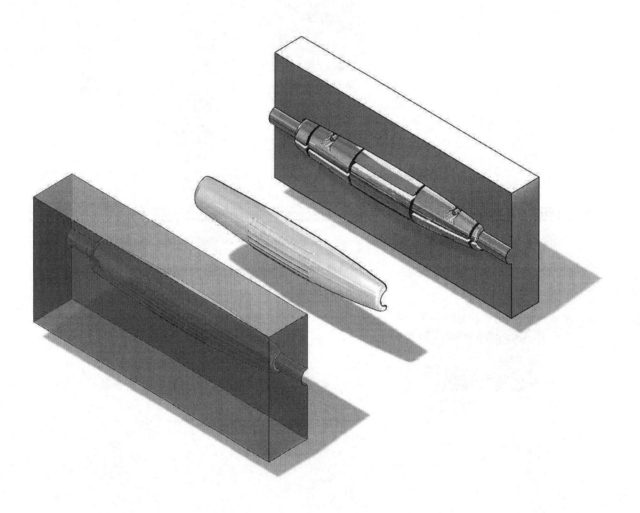

CHAPTER 2

Surface Repair

A lot of times, import files such as Parasolid, STEP, IGES, ACIS, and others may fail to produce solid or surface geometry because not all CAD systems support the same features, tolerances, or simply there are gaps and overlaps existing in the model.

This lesson will teach us some of the methods to repair the errors found in a surface model as well as converting it into a solid part.

1. Opening a Parasolid document:

Select **File, Open**.

Open a Parasolid document named: **Mouse.x_b**

2. Running Import Diagnostics:

The Import Diagnostics dialog appears when a non-SOLIDWORKS native document is opened.

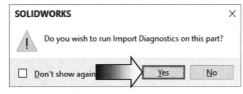

Import diagnostics repairs faulty surfaces, heals gaps between surfaces, knits repaired surfaces into closed bodies.
(Note: 3D Interconnect should be disabled for this lesson).

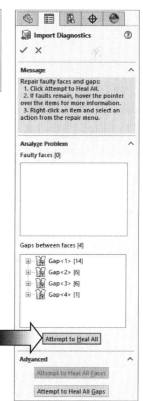

Click **YES** to run the Import Diagnostics utility.

There are **4 gaps** found in the model. They are displayed under the Gap Between Faces section.

Click **Attempt to Heal All** (arrow).

Both options Heal Gaps and Remove Gaps do not produce the desired results. We will repair the gaps and overlaps manually instead.

Click **Cancel** ☒.

3. Examining the small surfaces:

There are small slivers existing in the model. Zoom in on the t-shaped slot on top of the surface model.

There is a small triangular surface and a rectangular cutout in this area.

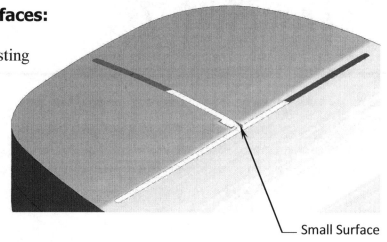

Small Surface

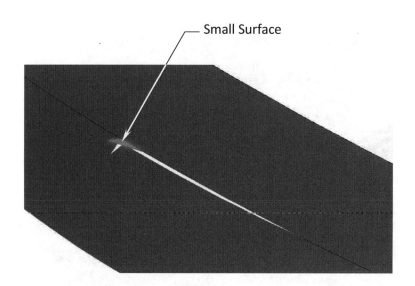

Small Surface

On the left side of the model, there is a gap and a small triangular surface in this area.

On the right end of the model, there is also a small triangular surface that did not get trimmed correctly.

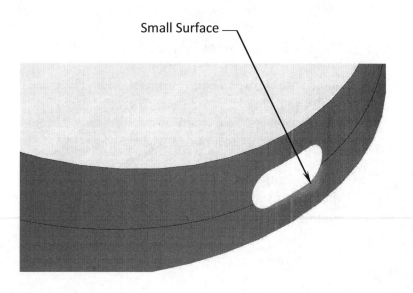

Small Surface

4. Deleting surfaces:

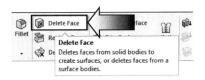

First, we will delete the 3 small triangular surfaces.

Change to the Surfaces tab and click **Delete Face** (arrow).

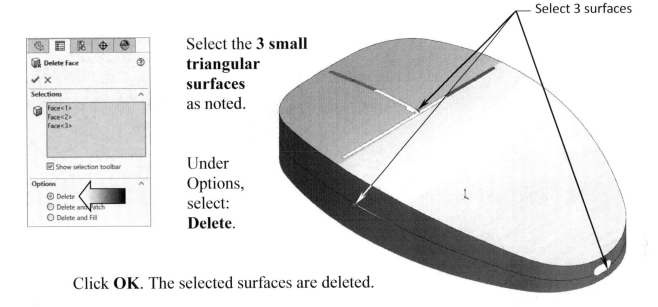

Select the **3 small triangular surfaces** as noted.

Under Options, select: **Delete**.

Click **OK**. The selected surfaces are deleted.

5. Deleting hole:

Next, we will delete the t-shaped lot on the top of the surface model.

Click **Delete Hole** on the Surfaces tab (arrow).

Under Selections, select the **13 edges** as indicated. (Right-click an edge and pick: **Select-Tangency** would be a little quicker.)

Click **OK**.

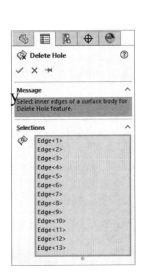

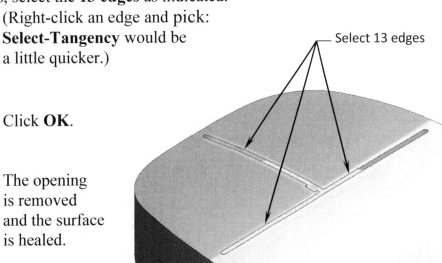

The opening is removed and the surface is healed.

6. Patching the openings:

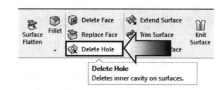

Zoom in on the gap on the <u>left side</u> of the surface model.

Click **Delete Hole** (arrow).

Select the **5 edges** of the opening.

Click **OK**.

The gap is healed.

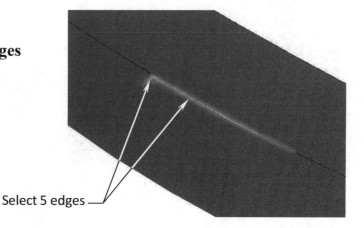

Select 5 edges

Zoom in on the <u>right end</u> of the model.

Click **Delete Hole** again.

Select the **5 edges** of the elliptical opening as noted.

Click **OK**.

The opening is removed.

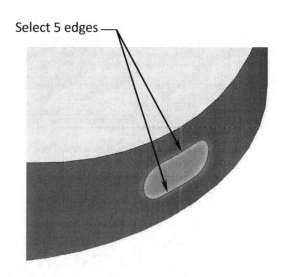

Select 5 edges

The gaps and openings are removed from the surface model.

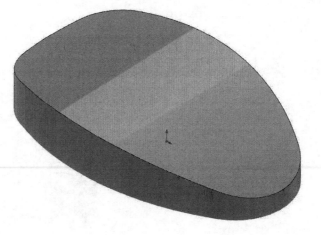

7. Adding draft:

A plastic injection molded part must have adequate drafts applied to all surfaces so that it can be ejected from the mold easily.

The surfaces on the sides of this surface model do not have any drafts at this point. We will add a 3 degrees draft to those surfaces now.

Change to the **Mold Tools** tab.

Click **Draft** (arrow).

For Type of Draft, select: **Neutral Plane**.

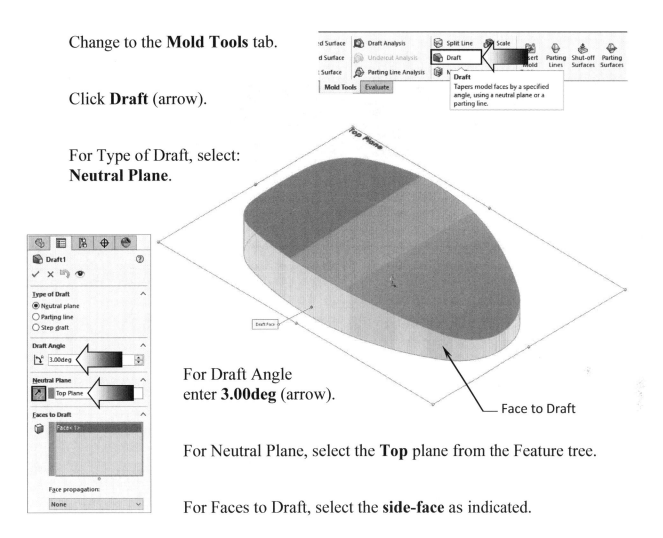

For Draft Angle enter **3.00deg** (arrow).

Face to Draft

For Neutral Plane, select the **Top** plane from the Feature tree.

For Faces to Draft, select the **side-face** as indicated.

Click **OK**.

Change to the Front orientation to view the draft angle on the side of the model.

3° Draft

3° Draft

8. Creating a split line feature:

The Split Line tool splits a surface into 2 surfaces so that each surface can be worked on individually.

Select the **Front** plane and open a <u>new sketch</u>.

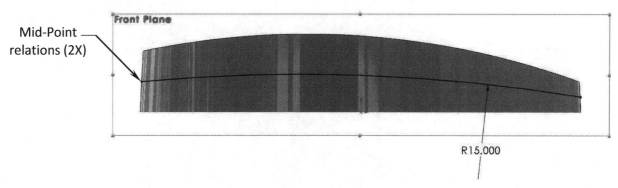

Sketch a **3-Point-Arc** and add the **Mid-Point** relations as indicated.

Add the **R15.000** dimension to fully define the sketch.

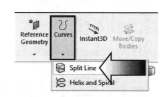

Change to the **Surfaces** tab and select: **Curves, Split Line** (arrow).

Use the default **Projection** type.

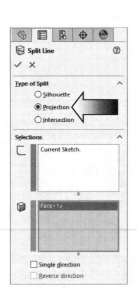

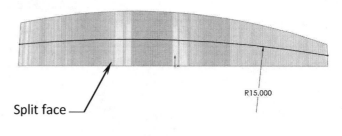

The **Current-Sketch** is selected automatically as the Split Sketch.

For Split Face, select the **side-face** of the model as indicated.

Click **OK**.

9. Creating a face fillet:

The Split Line created in the last step will be used as a boundary (or Hold Line) to determine the face fillet shape. The radius of the fillet is driven by the distance between the hold line and the edge to fillet.

Click **Fillet**.

For Fillet Type, select **Face Fillet**.

For Face Set 1, select the **top face**.

Select 2 edges for Hold Line

For Face Set 2, select the **side-face** as noted.

For Fillet Parameters, select **Hold Line** from the drop-down list.

For Hold Line, select the **2 edges** of the split line.

Leave all other parameters at their defaults.

Click **OK**.

The fillet shape and size were determined by the hold line (or the boundary of the split line). The fillet on the right end is smaller than the one on the left.

10. Adding thickness:

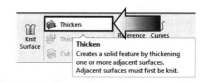

The surface model can now be thickened into a solid model so that other features can be added to it.

To rotate the model to the same orientation as the one shown below, click the following:

> * **Control + 7 = Isometric View**, and then:
> * **Shift + Up Arrow Key Twice = Reverse Isometric View**.

Click **Thicken** (arrow) on the Surfaces tab.

For Thicken Parameter, select the **Surface Model**.

For Thicken Direction, select: **Inside** (arrow).

For Thickness, enter **.070in.**

Click **OK**.

The surface model is thickened into a solid model, and a thickness of .070" is added to the inside wall.

11. Creating a recess cut:

The recess feature helps align the two plastic parts and also enhances their overall appearances.

Select the Top plane and open a **new sketch**.

Select edge to offset

Select the outer edge as noted and click: **Offset Entities**.

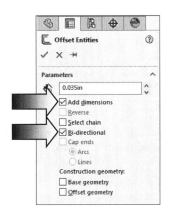

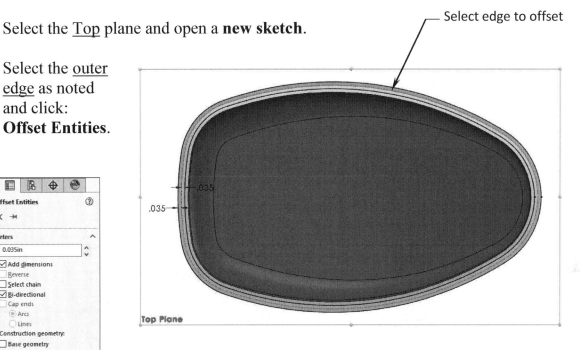

Enter **.035in** for offset Distance.

Enable the checkboxes: **Add Dimensions** and **Bi-Directional** (arrow), click **OK**.

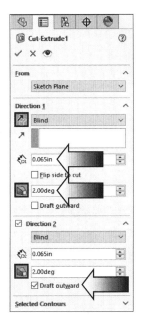

Change to the **Features** tab and click **Extruded Cut**.

Use the default **Blind** type.

Click **Reverse** and enter .065in. for depth.

Enable the **Draft** button and enter **2.00deg**.

Enable the **Direction 2** checkbox and apply the same parameters as the first.
Click **Draft Outward** for the direction 2 and click **OK**.

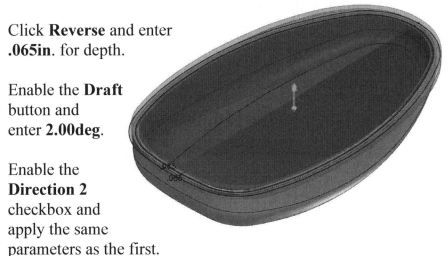

The recess cut feature is created. It is .065 inches deep and has a 2 degrees draft all around its perimeter.

Recess cut

Rotate the model to different orientations to inspect the recess cut feature.

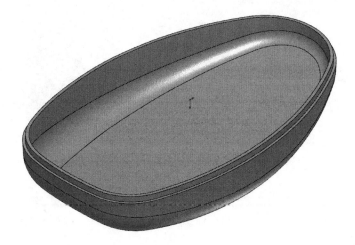

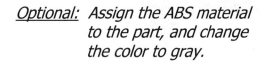
Optional: *Assign the ABS material to the part, and change the color to gray.*

12. Saving your work:

Select **File, Save As**.

Enter: **Mouse_Completed.sldprt** for the file name.

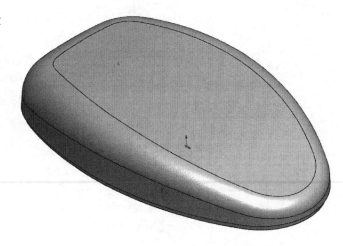

Click **Save**.

Close all documents.

CHAPTER 3

Core and Cavity Creation

There are several different methods
available to create the core and cavity blocks
for a mold; one of them is to create a tooling split.
The Engineered part must have at least three surface bodies in the Surface Bodies folder:
a **core** surface body, a **cavity** surface body, and a **parting** surface body.
This lesson will teach us one of the easier techniques to create the mold tooling for the part
shown below.

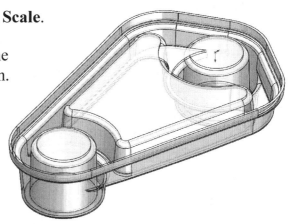

1. Opening a part document:

Select **File, Open**.

Locate the training files folder and
open a part document named:
Core and Cavity.sldprt

The material **PE High Density** has already been assigned
to this part. We will refer to the part as the Engineered Part.

2. Applying the scale factor:

The Scale feature scales only the geometry of the model for use in cavities to
accommodate material shrinkage. It does not scale dimensions, sketches, or
reference geometry. The scale formula: Cavity = part size *(1 + scale factor/100).

Change to the **Mold Tools** tab and click **Scale**.

For Scale About, use the
default **Centroid** option.

Enter **1.02** for
Scale Factor.

Click **OK**.

3. Creating the parting lines:

The Parting Lines run along the edges of the plastic part, between the core (green color) and the cavity surfaces (red color). They are used to create the parting surfaces, and to separate the surfaces.

Switch to the **Mold Tools** tab and click **Parting Lines** (arrow).

For Direction of Pull under Mold Parameters, select the **Top** plane from the FeatureManager tree.

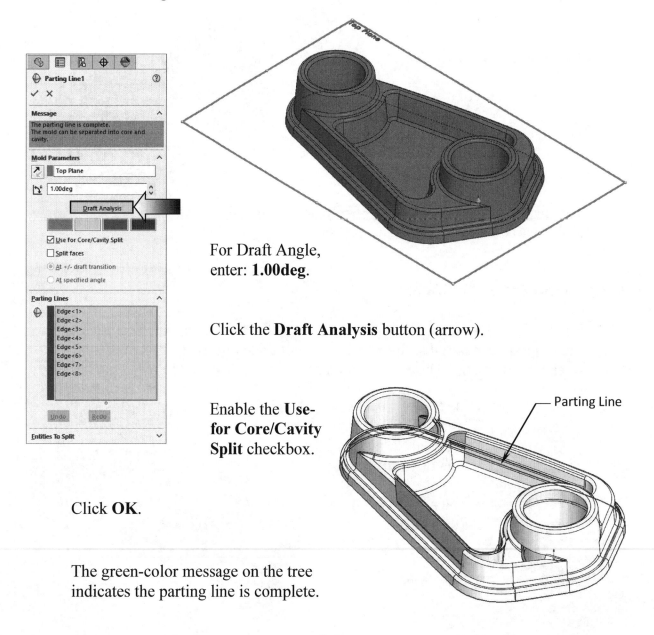

For Draft Angle, enter: **1.00deg**.

Click the **Draft Analysis** button (arrow).

Enable the **Use-for Core/Cavity Split** checkbox.

Parting Line

Click **OK**.

The green-color message on the tree indicates the parting line is complete.

4. Creating a parting surface:

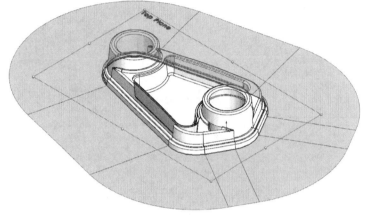

The Shut-off surface command closes up
the through holes in the part to create a core
surface and a cavity surface. Since there are no through
holes in this part, we can move on to creating the parting surfaces.

The parting surfaces are used to split the mold cavity from the core.

Click **Parting Surfaces** (arrow).

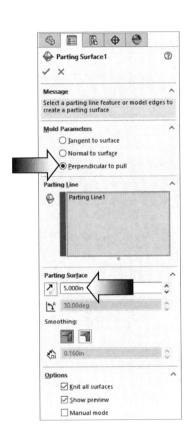

For Mold parameters,
select **Perpendicular to Pull** (arrow).

Parting Line1 is selected automatically.

For Parting Surface Distance, enter **5.000in** (arrow).

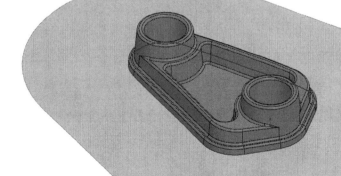

Leave the other
parameters at
their defaults.

Click **OK**.

The resulting Parting Surface is created.

5. Making the mold block sketch:

In order to create a tooling split, the part must have at least three surface bodies in the Surface Bodies folder: a core surface body, a cavity surface body, and a parting surface body.

A sketch is created to determine the final shape of the mold blocks and the block sizes are entered at the time of the tooling split.

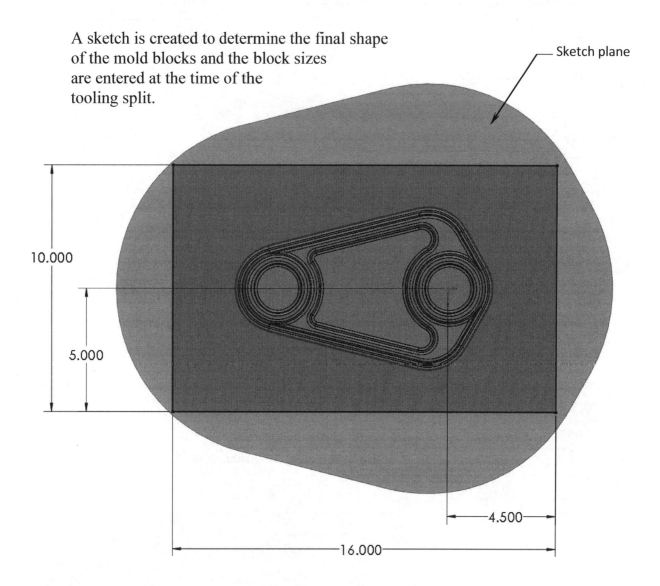

Select the Parting Surface as noted and open a **new sketch**.

Sketch a **Corner Rectangle** and add the dimensions as shown in the image.

Exit the sketch when it becomes fully defined.

6. Creating a tooling split feature:

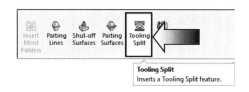

The last step is to use Tooling Split to create the core and cavity blocks for a mold.

Select the sketch of the **Rectangle** and click **Tooling Split** (arrow).

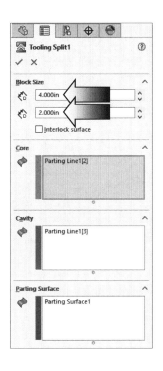

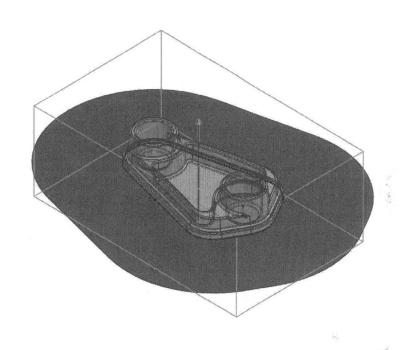

Under Block Size, enter the following for thickness of the two blocks:

Upper block: 4.000in.

Lower block: 2.000in.

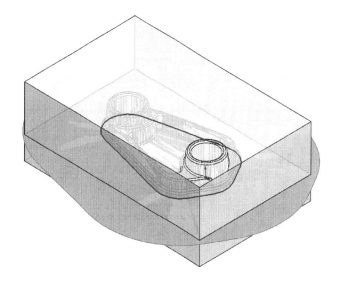

The graphics area displays the preview of the 2 mold blocks that are being created. (The image on the right was changed to transparent for clarity only.)

Click **OK**.

7. Hiding the references:

The references such as Parting Lines, Parting Surfaces should be hidden at this point so that the details of the 2 mold halves can be viewed more clearly.

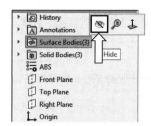

Select the **Surface Bodies** folder and click **Hide** (arrow).

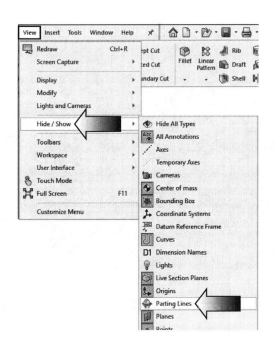

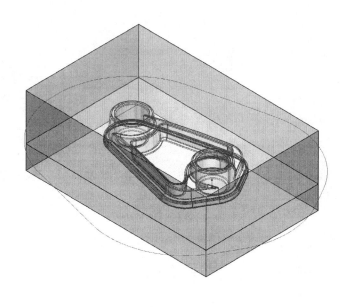

To hide the Parting Lines select: **View, Hide/Show, Parting Lines** (arrow).

Expand the **Solid Bodies** folder and
<u>rename</u> the 3 solid bodies to:

Plastic Part

Cavity Block

Core Block

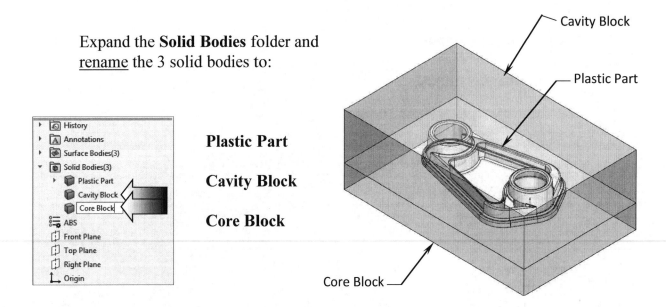

Cavity Block

Plastic Part

Core Block

8. Assigning materials:

Right-click the **Core Block** and select **Material**.

Select **Plain Carbon Steel** (arrow) from the drop-down list.

Assign the <u>same material</u> for the **Cavity Block**.

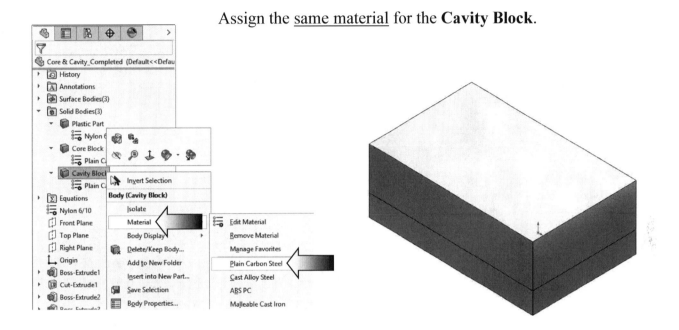

9. Changing the transparency:

Changing the transparency of the blocks will make it easier to view the interior details. The default transparency amount is 50%.

Right-click the **Cavity Block** and select **Change Transparency** (arrow).

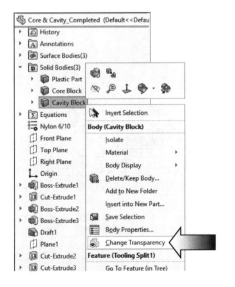

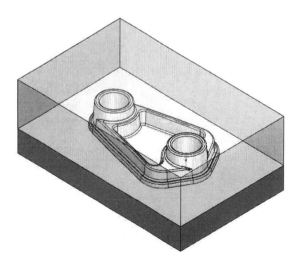

10. Separating the mold blocks:

The Exploded view command can be used to separate the blocks, but editing ability is disabled when the exploded view is active.

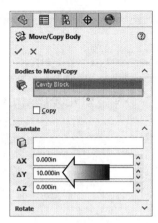

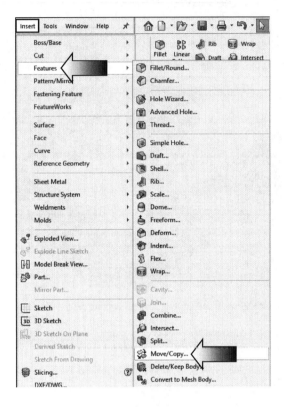

We will use an alternative method to separate the two mold blocks.

Select **Insert, Features, Move/Copy** (arrow).

Click the **upper block** (the cavity) and drag the **Y arrowhead** upward, approx **10.000in**.

Next, click the **lower block** (the Core) and drag the Y arrowhead downward, approximately **-10.000in**.

Click **OK**.

(The spacing of the two mold blocks is not critical. Either enter the spacing dimensions or simply drag the arrowheads to the desired distances.)

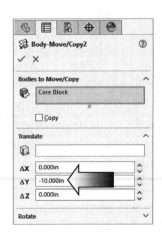

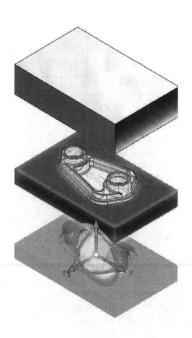

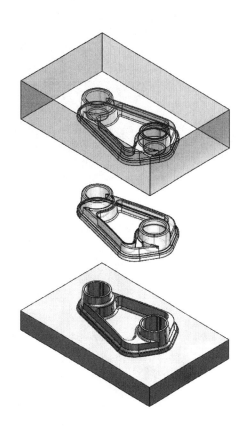

The Transparency of the Cavity Block may get reset after the exploded view is created.

Locate the Cavity Block under the Solid Bodies folder and change it to Transparent again, if needed.

11. Saving your work:

Select **File, Save As**.

Enter: **Core & Cavity Creation.sldprt** for the file name.

Click **Save**.

Close all documents.

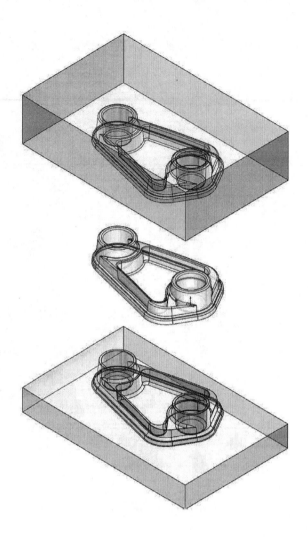

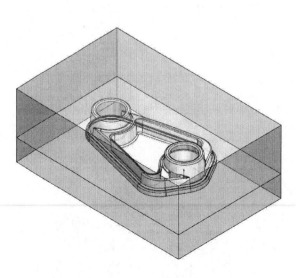

Exercise – Creating a Core and Cavity

The exercises are designed to encourage the students to apply the material learned from the lessons. These instructions are lean so as to give us a chance to try things out.

1. Opening a part document:

Select **File, Open**.

Open a part document named:
Core & Cavity_Exe.sldprt

The material **ABS** has already been assigned to this part.

2. Applying the scale:

Change to the **Mold Tools** tab.

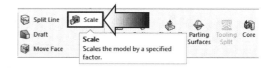

Click **Scale** (arrow). Press Shift + Up Arrow twice to rotate to the bottom isometric.

For Scale About, use the **Centroid** option.

Enable the **Uniform-Scaling** checkbox.

For Scale Factor, enter **1.015** (1.5% larger).

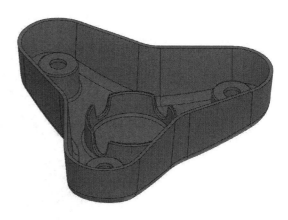

Click **OK**.

3. Creating the parting lines:

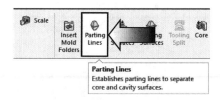

Click **Parting Lines** (arrow).

For Direction of Pull, select the **Top** plane from the FeatureManager tree.

For Draft angle, enter **1.00deg**.

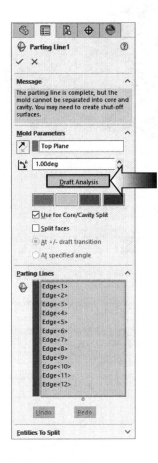

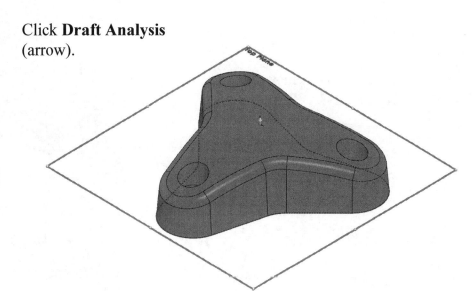

Click **Draft Analysis** (arrow).

The bottom edges are selected and used as the parting lines for this part.

Click **OK**.

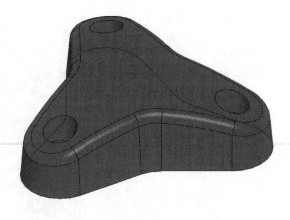

4. Creating the shut-off surfaces:

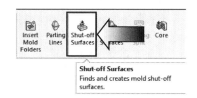

The 3 through holes in the part must be closed off before the parting surfaces can be added.

Click **Shut-Off Surfaces** (arrow).

The **3 edges** of the 3 holes are selected automatically.

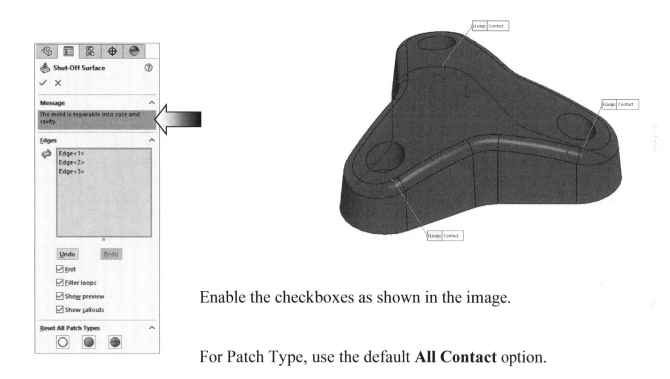

Enable the checkboxes as shown in the image.

For Patch Type, use the default **All Contact** option.

Click **OK**.

The green-color message indicates the mold is separable into Core and Cavity (arrow).

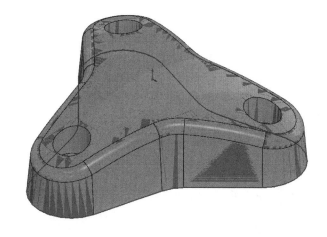

Expand the Surface Bodies folder to view the red and green surfaces. (They represent the Cavity surface and the Core surface.)

5. Creating the Parting Surfaces:

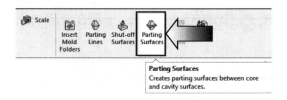

Change to the **Mold Tools** tab.

Click **Parting Surfaces** (arrow).

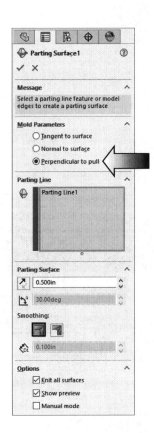

For Mold Parameters, select **Perpendicular to Pull** (arrow).

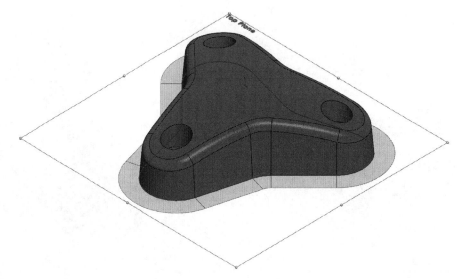

For Parting Surface Distance,
enter **.500in**.

For Smoothing, use the default **Sharp** option.

Enable the options:

Knit All Surfaces
Show Preview

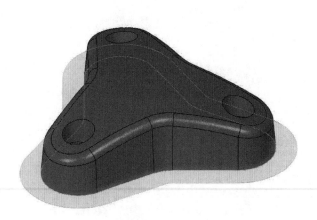

Click **OK**.

Expand the Surface Bodies
folder to view the Red surfaces (Core)
and the Green surfaces (Cavity).

6. Adding a new plane:

This plane is used to determine the interlock thickness. Change to the **Features** tab.

Select **Reference Geometry, Plane** (or select: **Insert, Features, Reference-Geometry, Plane**).

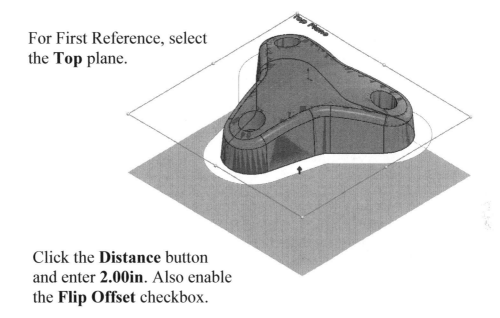

For First Reference, select the **Top** plane.

Click the **Distance** button and enter **2.00in**. Also enable the **Flip Offset** checkbox.

Click **OK**.

7. Making the sketch of the mold block:

Open a new sketch on the new Plane1.

Sketch a **Corner-Rectangle** centered on the origin.

Add dimensions.

Exit the sketch.

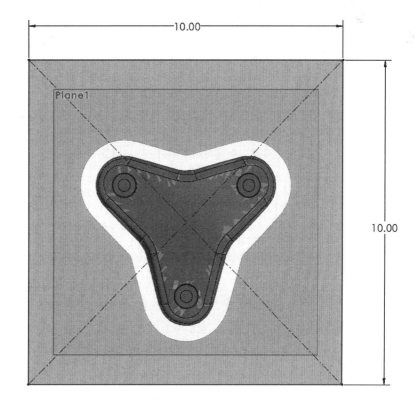

8. Inserting a Tooling Split feature:

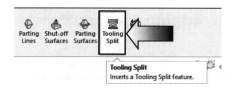

Switch to the **Mold Tools** tab.

Select the sketch of the **Rectangle** and click **Tooling Split** (arrow).

For Block Size, enter the following:

> * **Upper Block = 3.500in**.
>
> * **Lower Block -= 1.500in**.

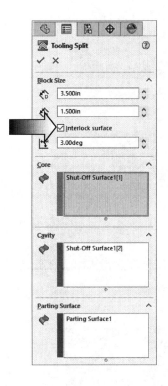

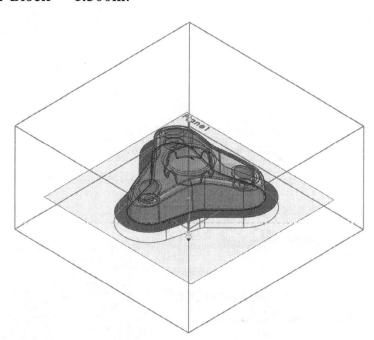

Enable the **Interlock Surface** checkbox.

Enter **3.00deg** for Draft Angle around the perimeter of the interlock surface.

Click **OK**.

The distance of the plane created in step number 6 defines the height of the interlock surface. The Interlock surface option will be discussed in the next few lessons.

9. Separating the mold blocks:

Select **Insert, Features, Move/Copy Bodies**.

Use the distance of **10.00in** along the <u>Delta Y</u> direction to move the upper block.

Use the distance of **-7.500in** along the <u>Delta Y</u> direction to move the lower block.

<u>Hide</u> all reference **surfaces** and also the **Parting Line**.

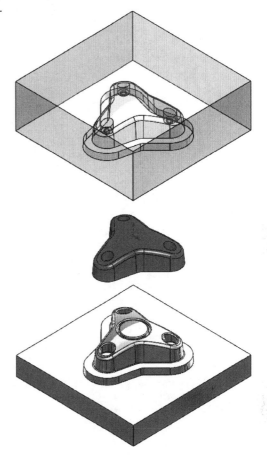

<u>Rename</u> the 3 solid bodies to:

        * **Plastic part**
        * **Cavity Block**
        * **Core Block**

Assign the material **Plain Carbon Steel** to the Cavity and the Core blocks.

Make the Cavity Block Transparent.

10. Saving your work:

Select **File, Save As**.

Enter: **Core & Cavity_Exe_Completed.sldprt** for the file name.

Click **Save**.

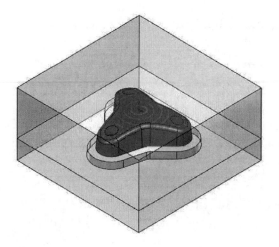

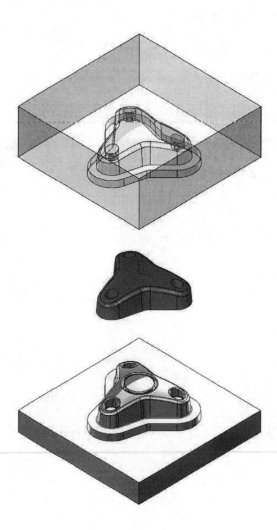

CHAPTER 4

Planar Parting Lines

The parting lines run along the edges of the plastic part, between the core (green surfaces) and the cavity surfaces (red surfaces). They are used to create the parting surfaces, and to separate the surfaces to make the 2 mold blocks. Multiple parting lines or partial parting lines can be created in a single part.

This lesson will teach us a method of creating the planar parting lines using a number of integrated tools that control the mold creation process.

1. Opening a part document:

Select **File, Open**.

Open a part document named:
Planar parting Lines.sldprt

The material **ABS** has already been assigned to this part.

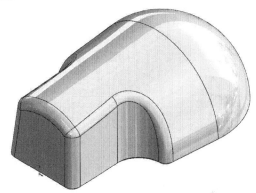

2. Scaling the part:

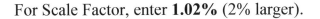

The scale tool will resize the plastic part to accommodate the shrink rate of the material but it does not change the dimensions, sketches, or any references of the part.

Change to the **Mold Tools** tab and click **Scale** (arrow).

For Scale About, select the **Centroid** option.

Enable the **Uniform-Scaling** checkbox.

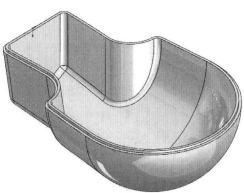

For Scale Factor, enter **1.02%** (2% larger).

Click **OK**.

3. Creating the parting lines:

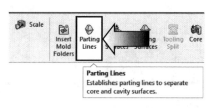

The parting line is one of the key features in a mold design. It is used to create the parting surfaces, and to separate the surfaces into a core and a cavity.

Click **Parting Lines** on the Mold Tools tab (arrow).

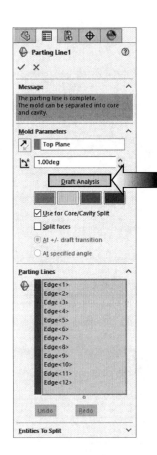

For Direction of Pull, select the **Top** plane from the Feature tree.

For Draft Angle, enter **1.00deg**.

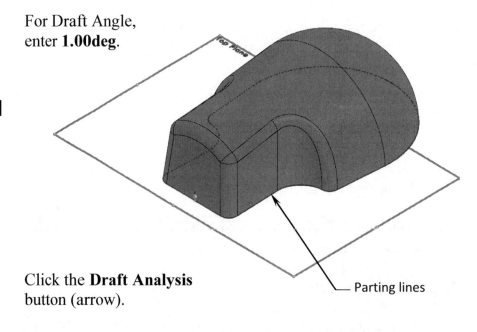

Click the **Draft Analysis** button (arrow).

Enable the checkbox: **Use for Core/Cavity Split**.

The bottom edges along the perimeter of the part are selected automatically, and a parting line is created.

Click **OK**.

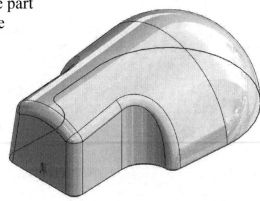

Rotate the model to see the parting line (in blue color) more easily.

4. Creating the parting surfaces:

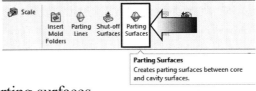

There is no through hole in the part;
we can skip the step of making the Shut-Off
surfaces, and move forward to making the parting surfaces.

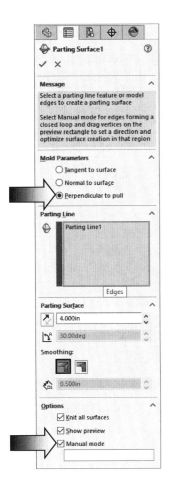

Click **Parting Surfaces** (arrow).

For Mold Parameters, select: **Perpendicular to Pull** (arrow).

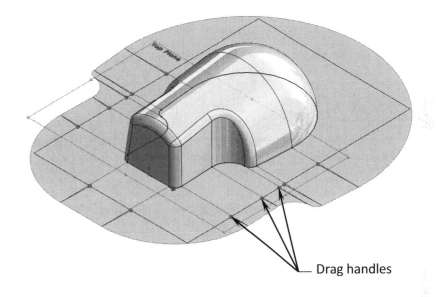

Drag handles

The system selects the **Parting Line1** automatically.

For Parting Surface Distance, enter **4.00in**.

For Smoothing, use the default **Sharp** option.

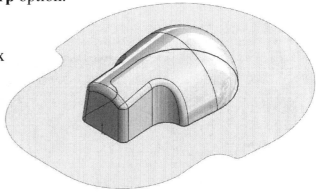

Enable the **Manual Mode** checkbox
and drag the handles to prevent
the intersection of overlapping
surfaces.

Click **OK**.

5. Making the mold block sketch:

The Tooling Split tool uses a sketch to define the shape and size of the mold insert.

Select the <u>parting surface</u> and open a **new sketch**.

Sketch a **Corner Rectangle** and add the dimensions shown in the image to fully define the sketch.

<u>Exit</u> the sketch.

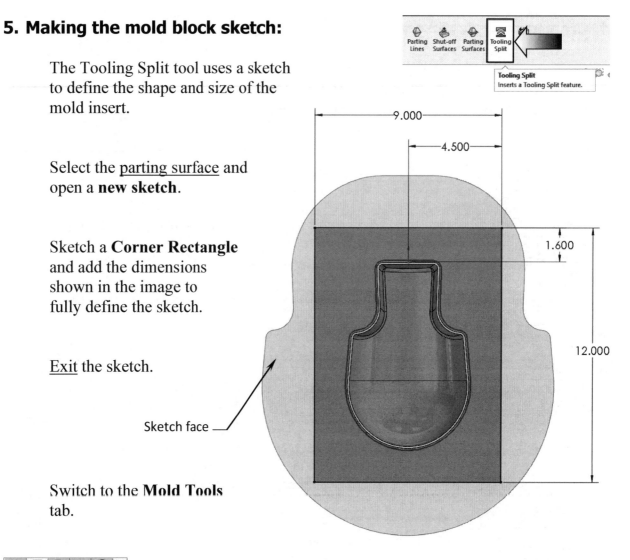

Sketch face

Switch to the **Mold Tools** tab.

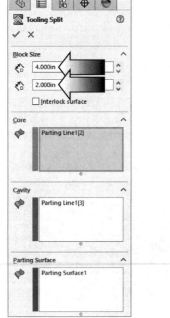

Select the sketch of the **rectangle** and click **Tooling Split**.

For Block Size, enter:

4.000in.
2.000in.

Click **OK**.

The 2 mold blocks are created.

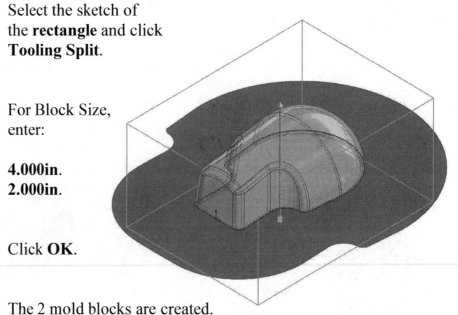

6. Adding the Move/Copy command:

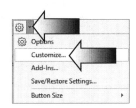

The Move/Copy Bodies command will be used many times throughout this textbook. Let us add the icon to the Features toolbar instead of using the drop-down menus frequently.

Select **Tools, Customize** and click the **Commands** tab (arrow).

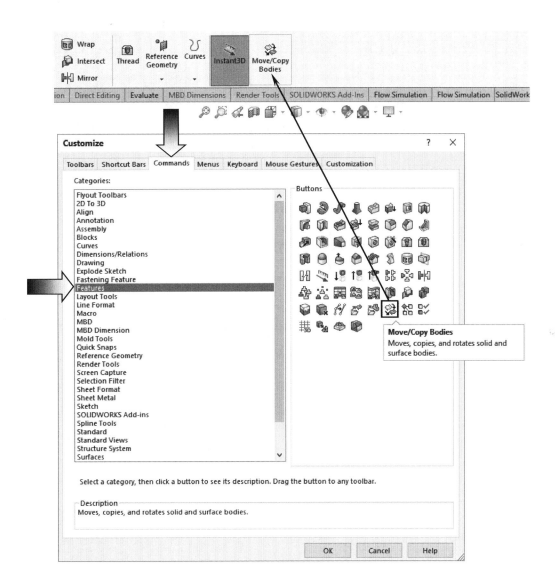

Under **Categories**, click **Features** (arrow) and <u>drag</u> the **Move/Copy Bodies** button to the Features toolbar, on far right as indicated.

Click **OK** to close the Customize dialog box.

7. Separating the mold blocks:

The Move/Copy Bodies tool can move, rotate, and copy solid and surface bodies or place them using mates.

Click **Move/Copy Bodies**.

For Bodies to Move/Copy, select the **upper block** (the Cavity).

Click in the <u>Delta Y</u> field and enter **10.00in** (arrow).

Click **OK**.

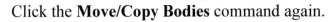

The Cavity block moves upwards 10 inches.

Click the **Move/Copy Bodies** command again.

For Bodies to Move/Copy, select the **lower block** (the Core).

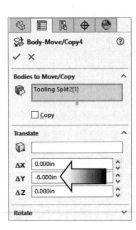

Click in the <u>Delta Y</u> field and enter **-8.00in**.

Click **OK**.

The Core block moves downward 8 inches. The references will be put away in the next couple of steps.

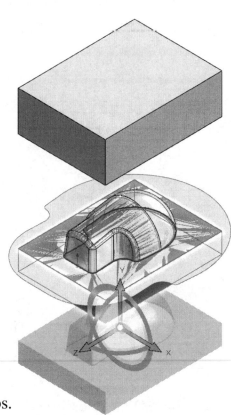

8. Hiding the references:

The references such as parting lines and parting surfaces will need to be hidden at this point so that the details of the 2 mold halves can be viewed more clearly.

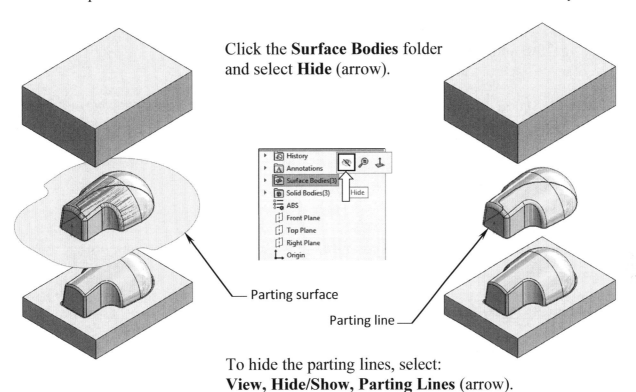

Click the **Surface Bodies** folder and select **Hide** (arrow).

— Parting surface

Parting line —

To hide the parting lines, select:
View, Hide/Show, Parting Lines (arrow).

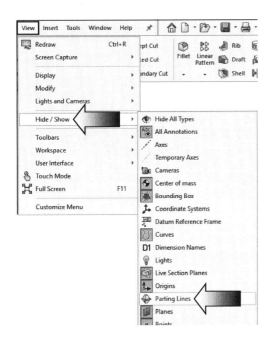

To complete the mold each solid body will be saved individually as part files and then inserted into an assembly document.

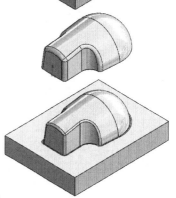

The mold blocks are joined to the mold base and the remaining details such as gates, runners, cooling lines are added at that point.

All 3 solid bodies have the same color at this point, which makes it a little bit confusing. We will change it in the next couple of steps.

9. Renaming the mold blocks:

Renaming the solid bodies to reflect what they represent is highly recommended.

Once renamed, the solid bodies can easily be identified, making future editing a little quicker.

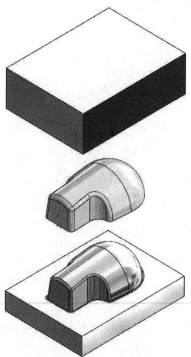

Slow double-click on the 1st solid body (or press F2) and rename it to **Plastic Part**.

Rename the 2nd solid body to **Cavity Block** (arrow).

Rename the 3rd solid body to **Core Block** (arrow).

10. Assigning the materials:

Each solid body can have its own material assigned to it.

Right-click the **Cavity Block**, click **Material**, and select: **Plain Carbon Steel** (arrow).

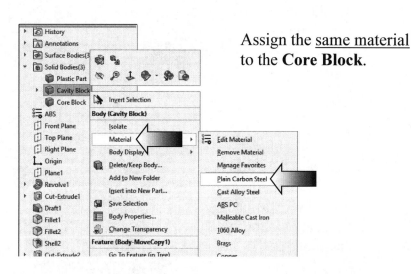

Assign the same material to the **Core Block**.

The Plastic Part already has the ABS material assigned to it.

11. Changing to transparency:

The details of the Cavity Block can be viewed more clearly when changed to transparent.

Right-click the Cavity Block and select: **Change Transparency** (arrow).

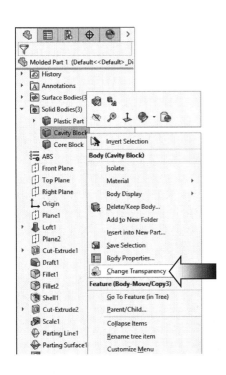

The Cavity Block is changed to transparent.

The default transparent amount is 75%. This setting can be modified by selecting:

* The Cavity Block
* Appearances
* Body
* Advanced
* Illumination
* Transparent Amount

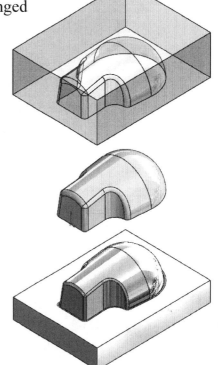

Note: the lessons in this textbook will only focus on the key steps to designing the core and cavity mold blocks such as Scale, Planar and Non-Planar Parting Lines, Shut-Off Surfaces, Parting Surfaces, Tooling Split, and Interlock Surfaces.

12. Saving your work:

Select **File, Save As**.

Enter **Planar Parting Lines_Completed.sldprt** for the file name.

Click **Save**.

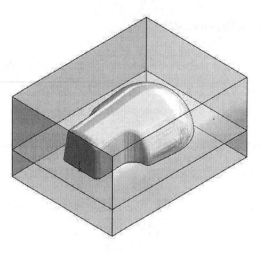

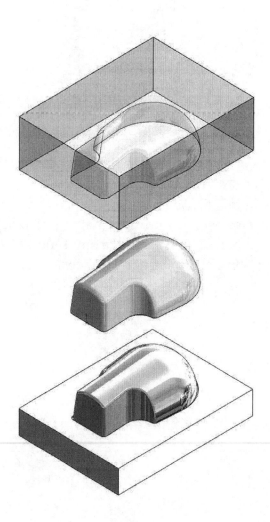

Exercise – Planar Parting Lines

Parting lines are created from the edges of a plastic part. They are used to separate the surfaces that belong to the core and cavity. They are also the edges that form the inside perimeter of the parting surfaces.

When the parting lines are being created, the Draft Analysis tool is used to identify the positive side (green surfaces) and the negative side (red surfaces), and later, the parting surfaces will be created from them.

1. Opening a part document:

Select **File, Open**.

Open a part document named:
Planar Parting Lines_Exe.sldprt

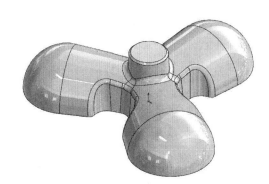

The material **ABS** has already been assigned to this part.

2. Scaling the part:

Switch to the **Mold Tools** tab.

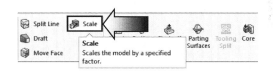

Click **Scale** (arrow) and select the **Centroid** option.

Enable the **Uniform Scaling** checkbox.

For Scale Factor, enter
1.02 (2% larger).

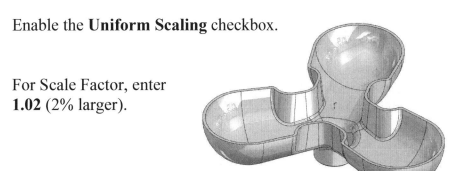

Click **OK**.

3. Creating the parting lines:

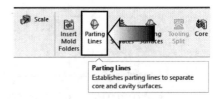

Select the **Parting Lines** command (arrow).

For Direction of Pull, select the **Top** plane from the FeatureManager tree.

For Draft Angle, enter **1.00deg**.

Click **Draft Analysis** (arrow).

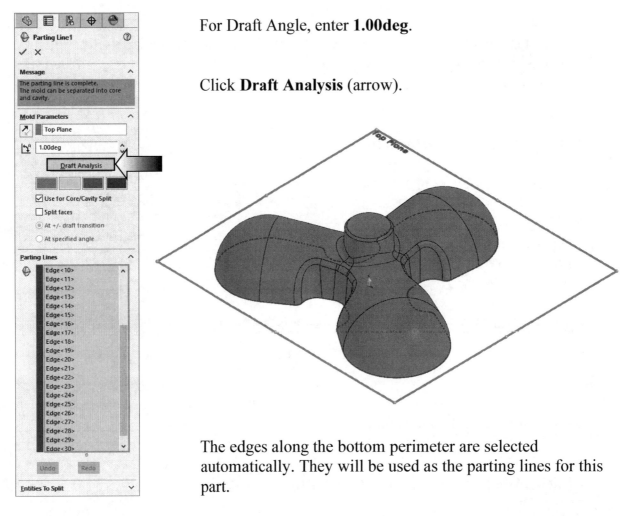

The edges along the bottom perimeter are selected automatically. They will be used as the parting lines for this part.

Click **OK**.

Keep all other parameters at their defaults.

A parting line is created and displayed in blue color.

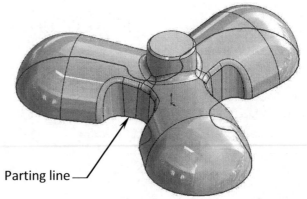

Parting line

4. Creating the parting surfaces:

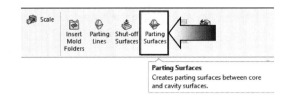

Click **Parting Surfaces** (arrow).

For Mold Parameter, select **Perpendicular to Pull** (arrow).

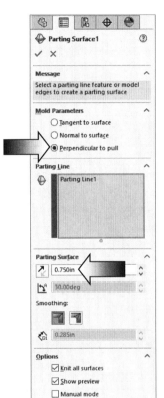

The **Parting Line1** is selected automatically.

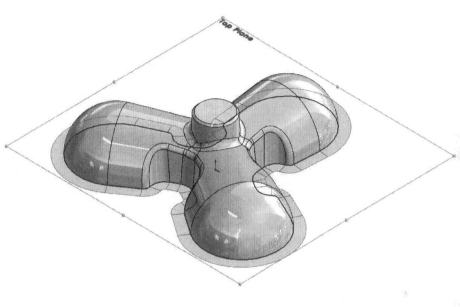

For parting Surface Distance, enter **.750in**. (arrow).

For Smoothing, use the default **Sharp** option.

Enable the **Knit** and **Show Preview** checkboxes. (The Manual Mode will display the handles to manipulate and adjust the parting surface when needed.)

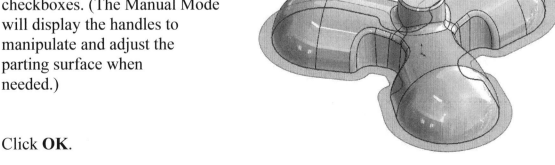

Click **OK**.

5. Adding a new plane:

This new plane is used to sketch the shape of of the mold blocks and also to set the height of the interlock surface.

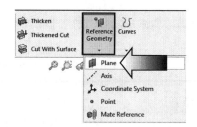

Change to the **Surfaces** tab and select: **Reference Geometry, Plane**.

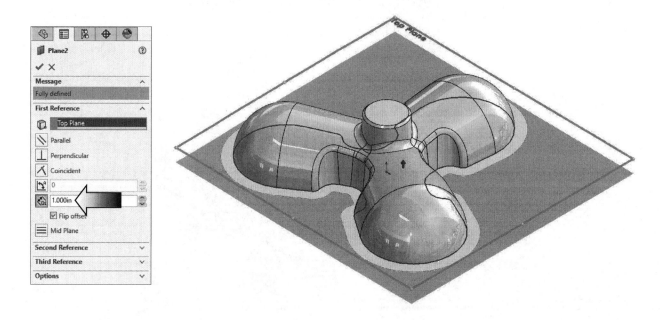

For First Reference, select the **Top** Plane from the FeatureManager tree.

Enable the Offset Distance button and enter **1.00in**. (arrow),

Click **Flip Offset** to place the new plane <u>below</u> the Top plane.

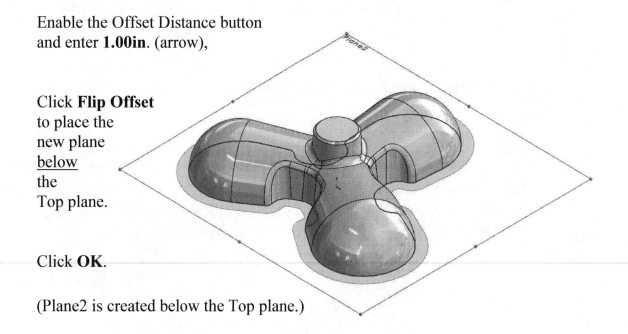

Click **OK**.

(Plane2 is created below the Top plane.)

6. Making the mold block sketch:

Select the new <u>Plane2</u> and open a **new sketch**.

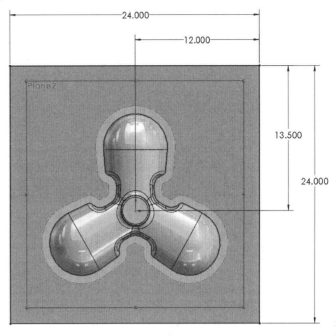

Sketch a **Corner Rectangle** and add dimensions to fully define the sketch.

<u>Exit</u> the sketch.

7. Inserting a tooling split:

Switch back to the **Mold Tools** tab.

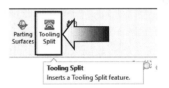

Select the sketch of the **Rectangle** and click **Tooling Split**.

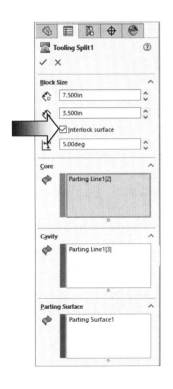

For Block Size, enter the following:

* **Upper Block = 7.500in**.

* **Lower Block = 3.500in**.

Enable the **Interlock-Surface** checkbox.

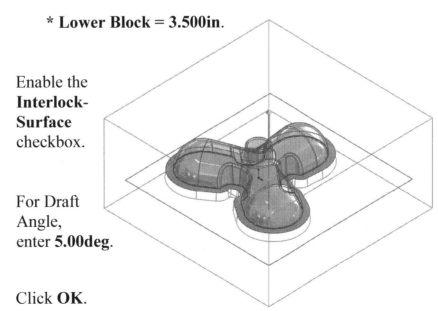

For Draft Angle, enter **5.00deg**.

Click **OK**.

8. Changing materials:

Separate the 2 blocks using the **Move/Copy** command.

To hide all surfaces, right-click the **Surface Bodies** folder and select **Hide**.

To hide the parting line, select: **View, Hide/Show, Parting Lines**.

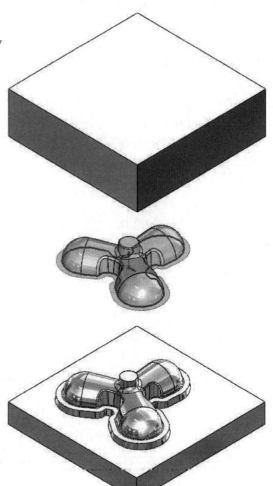

Rename the 3 solid bodies to:

Plastic Part

Cavity Block

Core Block

Right-click the Cavity Block (and the Core), select **Material**, **Plain Carbon Steel**.

Right-click the **Cavity Block** and select: **Change Transparency**.

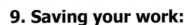

9. Saving your work:

Select **File, Save As**.

Enter: **Planar Parting Lines_Exe_ Completed.sldprt** for the file name.

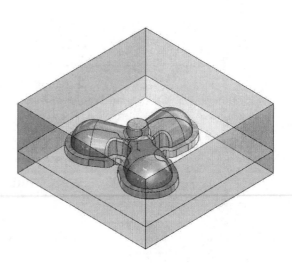

Click **Save**.

Close all documents.

CHAPTER 5

Interlock Surface 1

The Interlock Surfaces runs around the perimeter
of the parting surface. It usually has 3 to 5 degrees draft, away from the parting lines.
The interlock surface can help seal the mold properly to prevent plastic from leaking. It can
guide the tooling into place and maintain alignment during the molding process, and also
prevent shifts, uneven surfaces, or incorrect wall thicknesses.

1. Opening a part document:

Select **File, Open**.

Open a part document named:
Interlock Surface 1.sldprt

The material **ABS** has already been
assigned to the part.

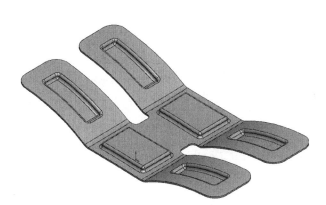

2. Applying the scale factor:

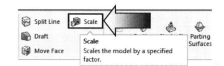

Switch to the **Mold Tools** tab and click **Scale**.

The Scale feature changes the overall geometry of the part, but it does not scale
dimensions, sketches, or reference geometry.

For Scale About, select the **Centroid** option.

Enable the **Uniform Scaling** checkbox.

For Scale Factor,
enter: **1.02** 2% larger).

Click **OK**.

The part is 2% larger than its original size to accommodate for shrinkage.

3. Creating the parting lines:

The parting lines run along the edge of the plastic part, between the core and the cavity surfaces. They are used to create the parting surfaces, and to separate the core and cavity surfaces.

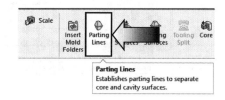

The parting lines for this model, due to its geometry, will be a non-planar parting line. The interlock surface may need to be created to help align the two mold-halves.

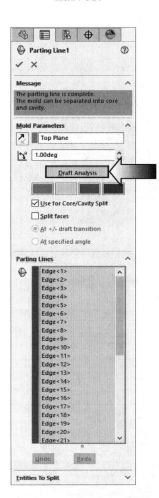

Click **Parting Lines**.

For Direction of Pull, select the **Top** plane from the FeatureManager tree.

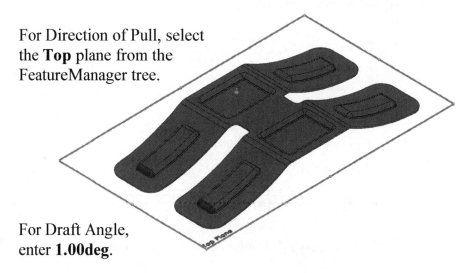

For Draft Angle, enter **1.00deg**.

Click the **Draft Analysis** button (arrow). (Since the part is very thin and flexible, we will ignore the straddle faces (blue) for this exercise).

A green-color message appears on the upper left side, it indicates the parting lines is completed.

Click **OK**.

4. Creating the parting surfaces:

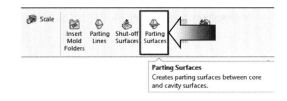

The Parting Surface is extended from the parting lines. It is used to split the mold cavity from the core.

Select the **Parting Surfaces** command (arrow) from the **Mold Tools** tab.

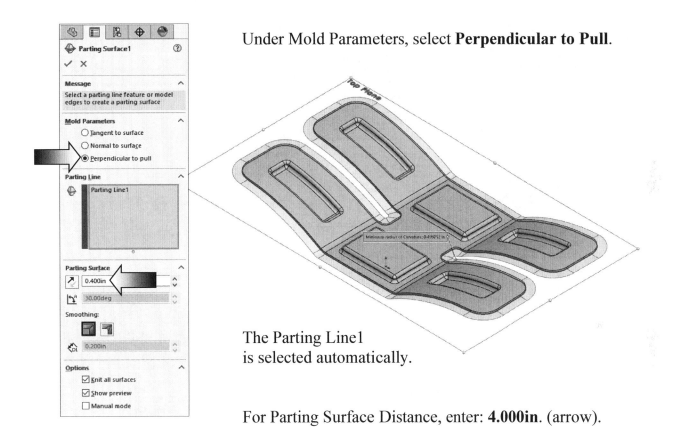

Under Mold Parameters, select **Perpendicular to Pull**.

The Parting Line1 is selected automatically.

For Parting Surface Distance, enter: **4.000in**. (arrow).

Keep all other parameters at their default settings.

Click **OK**.

A set of parting surfaces are created and protruded outward from the parting lines. They can overlap if the distance is too big.

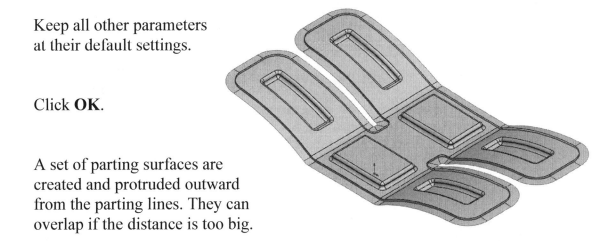

5. Making the mold block sketch:

The mold block sketch is used to define the shape and size of the mold blocks. It can be any shape such as rectangular, circular, or square.

Open a **new sketch** on the Top plane.

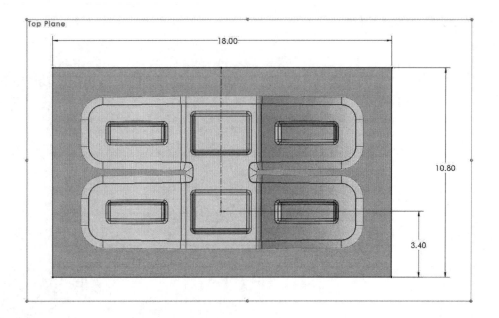

Sketch a **Corner Rectangle** and add the dimensions shown in the image.

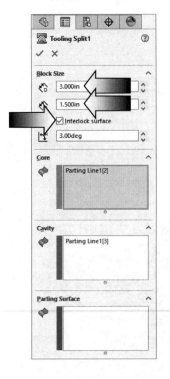

Exit the sketch. Click **Tooling Split** on the Mold Tools tab. (Select the sketch of the rectangle if prompted).

Under Block Size, enter the following:

3.000in
1.500in

Enable the **Interlock Surface** check-box and enter **3.00deg**.

Click **OK**.

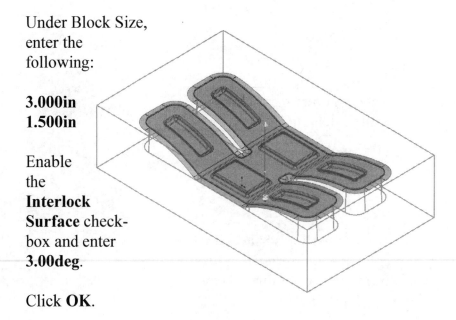

6. Hiding the references:

The reference surfaces and the parting lines are usually put away for clarity, after the core and cavity are created.

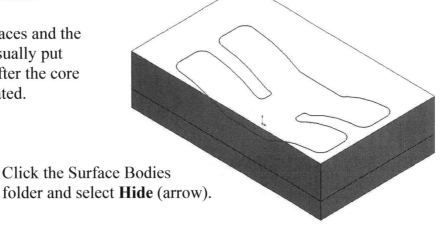

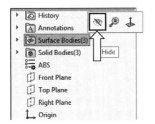

Click the Surface Bodies folder and select **Hide** (arrow).

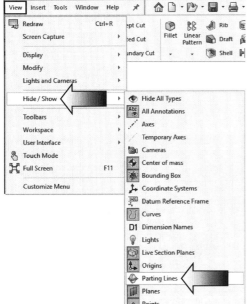

To hide the Parting lines, select:
View, Hide/Show, Parting Lines (arrow).

To rename the solid bodies, expand the Solid Bodies folder and rename the 3 solid bodies to:

* **Plastic Part**
* **Core Block**
* **Cavity Block**

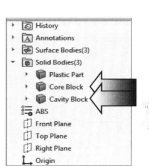

To assign material to the core and the cavity, right-click the **Core Block**, select: **Cast Alloy Steel**.

Assign the same material to the Cavity Block.

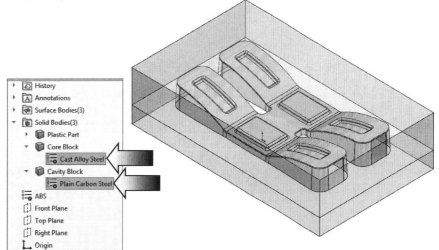

7. Separating the mold blocks:

The interior details are much easier to see when the mold blocks are separated.

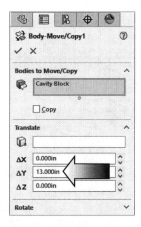

Select **Insert, Features, Move/Copy**.

For Body to Move/Copy, select the **Cavity Block**.

Click in the **Delta Y** field and enter **13.000in**.

Click **OK**.

The Cavity Block moves upward 13 inches.

Click **Insert, Features, Move/Copy** again.

For Body to Move/Copy, select the **Core Block**.

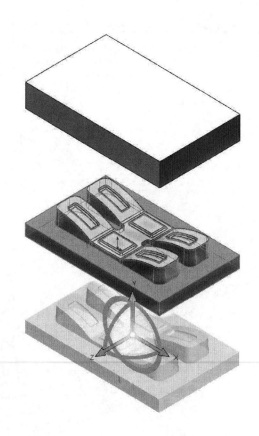

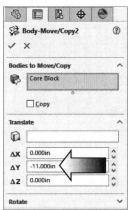

Click in the **Delta Y** field and enter **-11.000in**.

Click **OK**.

The Core Block moves downward **-11.000in**.

The Core and Cavity blocks are separated.

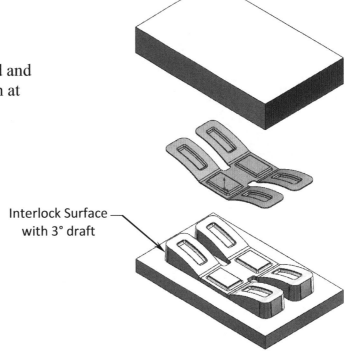

The 2 mold blocks are separated and the interlock surface can be seen at this point.

A 3 degrees draft was added to the side faces of the interlock surfaces during the tooling split operation.

Interlock Surface with 3° draft

8. Changing the transparency:

Expand the **Solid Bodies** folder.

Right-click on the Cavity Block and select: **Change Transparency** (arrow).

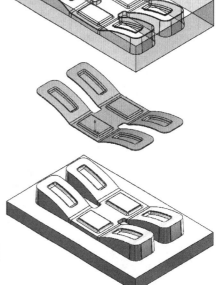

The details in the Cavity Block are now visible in the isometric view.

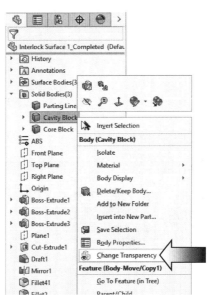

9. Saving your work:

Select **File, Save As**.

Enter: **Interlock Surface 1_Completed.sldprt** for the file name.

Click **Save**.

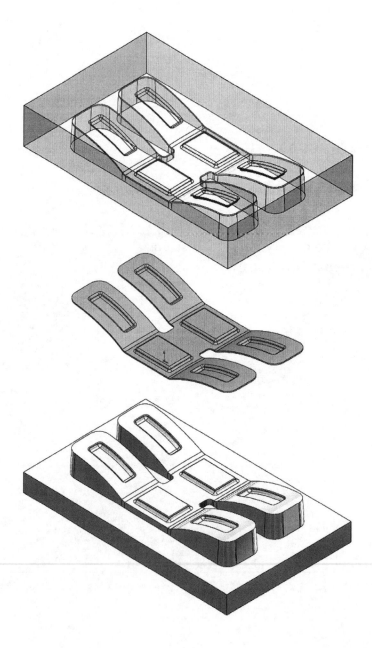

Interlock Surface 2

As mentioned in the 1st half of the lesson, the interlock surfaces can help seal the mold properly to prevent plastic from leaking. It can also guide the tooling into place and maintain alignment during the molding process. Additionally, interlock surface can prevent shifts, uneven surfaces, or incorrect wall thicknesses.

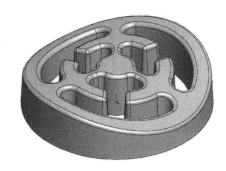

This lesson will walk us through the creation of a circular interlock surface for a cylindrical shaped plastic part.

1. Opening a part document:

Select **File, Open**.

Open a part document named: **Interlock Surface 2.sldprt**.

2. Applying the scale factor:

The material **ABS** has already been assigned to the part. The first step is to apply the scale factor to accommodate the shrink rate of this material.

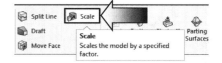

Change to the **Mold Tools** tab and click **Scale** (arrow).

For Scale About, select the **Centroid** option.

Enable the **Uniform Scaling** checkbox.

For Scale Factor, enter: **1.02** (2% larger).

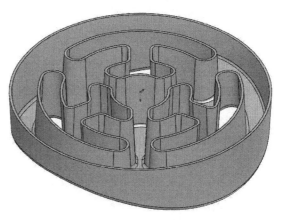

Click **OK**.

3. Creating the parting lines:

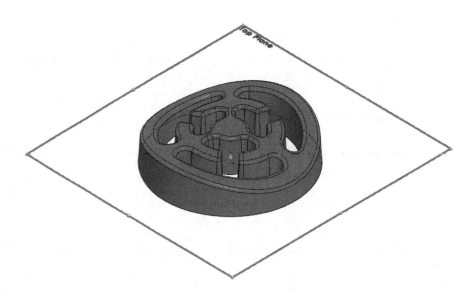

The parting lines run along the edge of the plastic part, between the core (green surfaces and the cavity (red surfaces). They are used to create the parting surfaces, and also to separate the core and cavity surfaces.

Parting Lines
Establishes parting lines to separate core and cavity surfaces.

Select the **Parting Lines** command on the Mold Tools tab.

Under the Mold Parameters, for Direction of Pull, select the **Top** plane from the FeatureManager tree.

For Draft Angle, enter **1.000deg**.

Click **Draft Analysis** (arrow).

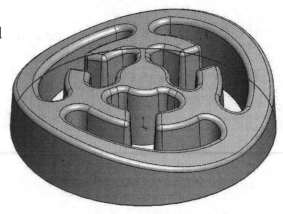

A parting line is created along the bottom edge of the part.

Keep other settings at their defaults.

Click **OK**.

4. Adding the shut-off surfaces:

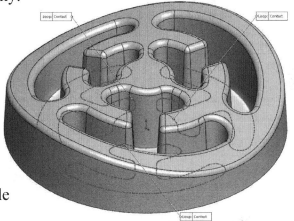

The Shut-off surfaces tool closes up the through holes in a plastic part. Only one Shut-off Surface feature is allowed in a part. A fill type of No Fill, Contact, or Tangent, must be assigned to every through hole.

Select the **Shut-Off Surfaces** command on the Mold Tools tab.

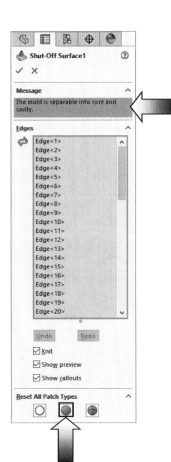

All edges in the model that formed the openings are closed off automatically.

A green message appears on the upper left of the Properties tree indicating that "The mold is separable into core and cavity".

Enable the checkboxes for the **Knit**, **Show Preview**, and **Show Callouts**.

For Patch Type, select the **All Contact** option (arrow),

Click **OK**.
(Expand the Surface Bodies folder to see the Red [Core] and Green [Cavity] surfaces).

5. Creating the parting surface:

The parting surfaces get created after the
parting lines and shut-off surfaces are made.
They are used to split the mold cavity from the core.

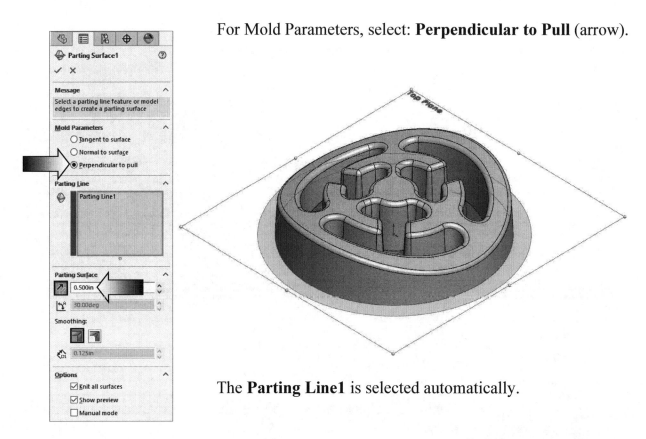

Click **Parting Surfaces** (arrow).

For Mold Parameters, select: **Perpendicular to Pull** (arrow).

The **Parting Line1** is selected automatically.

For Parting Surface Distance, enter **.500in**. and click **Reverse**.

For Smoothing, select the **Sharp** option.

Clear the **Manual Mode** checkbox
(This option displays handles to
adjust the parting surface).

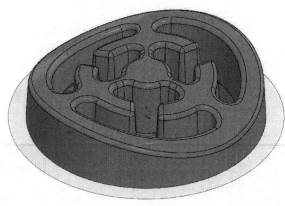

Click **OK**.

6. Adding a new plane:

Depends on the size of the part, the interlock surface is usually created from a plane that is offset below the parting lines. It helps guide the mold into place and prevents them from shifting during the mold process.

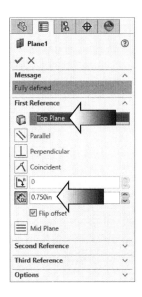

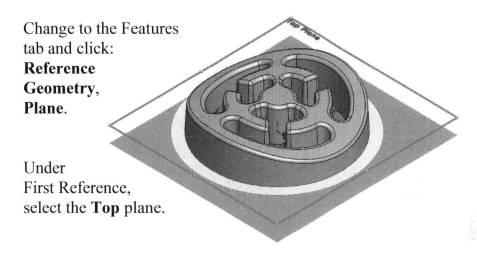

Change to the Features tab and click:
Reference Geometry, Plane.

Under First Reference, select the **Top** plane.

Enable the Distance button, enter **.750in** and click **Flip Offset** to place the plane <u>below</u> the Top plane.

Click **OK**.

7. Making the mold block sketch:

Select the <u>new plane</u> and open a **new sketch**.

Sketch a **Circle** centered on the origin.

Add a diameter dimension of **10.000in**.

Exit the sketch.

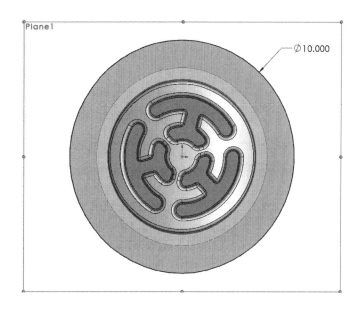

8. Creating a tooling split feature:

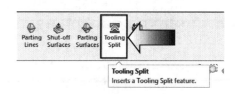

The Interlock surface also seals the liquid plastic from leaking and maintains the alignment between the core and cavity.
An interlock surface will be added to this tooling split.

Select the <u>sketch</u> from the last step and click **Tooling Split** (arrow).

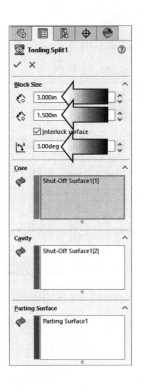

For Block Size, enter:

Upper Block = 3.000in.

Lower Block = 1.500in.

Enable the **Interlock Surface** checkbox.

Enter **3.00deg** for Draft angle.

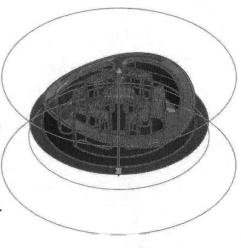

The reference surfaces for the Core, Cavity, and Parting Surface are automatically selected from the Surface Bodies folder.

The preview graphics shows the Core and Cavity blocks are being created.

Click **OK**.

The exploded view command can be used to separate the two mold blocks but all editing will be disabled while the exploded view is active. We will use the Move/Copy command to separate the core and cavity instead. That way editing can be done when the view is still exploded.

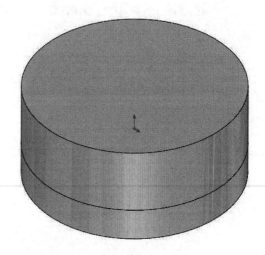

9. Separating the mold blocks:

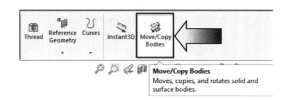

Select the **Move/Copy Bodies** command or select: **Insert, Features, Move/Copy**.

For Body to Move/Copy, select the **upper block** (the Cavity).

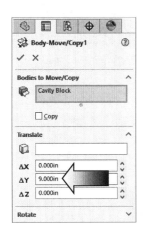

Under Translate, click in the <u>Delta Y</u> field and enter **9.000in**.

Click **OK**.

The upper block moves upwards 9 inches.

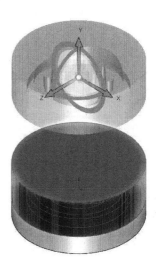

Select the **Move/Copy Bodies** command again or select: **Insert, Features, Move/Copy**.

For Body to Move/Copy, select the **lower block** (the Core).

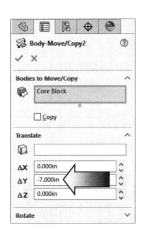

Under Translate, click in the <u>Delta Y</u> field and enter **-7.000in**.

Click **OK**.

The lower block moves downward 7 inches.

The distances between the two mold blocks are not critical.

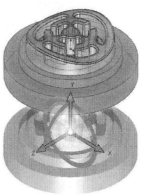

10. Hiding the references:

The references used to create the mold blocks should be hidden at this point.

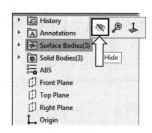

To hide all reference surfaces, click the **Surface Bodies** folder and select **Hide** (arrow).

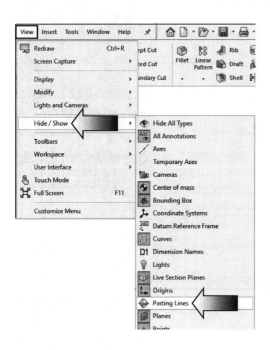

To hide the parting lines, select: **View, Hide/Show, Parting Lines** (arrow).

The interlock surface is now visible. The Core block has a raised feature added to it and the Cavity block has the matching recess feature.

Push **Control + 7** to change to the Isometric view, and **Shift + Up Arrow twice**, to change to the Reverse Isometric view. The recessed interlock surface can now be viewed more clearly.

Interlock Surface —

11. Renaming the solid bodies:

It will be easier for everyone to identify the solid bodies if their names were changed to reflect what they are.

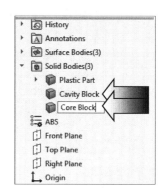

The solid bodies can also be saved as individual part files so that they can be shared with other team members. One way to save a solid body is to right-click on its name and select Insert into New Part and enter a name to save. The new part remains linked to the original document. Changes done to the original part document will populate to the new part instantly.

Expand the solid Bodies folder and <u>rename</u> the three solid bodies to:

Plastic Part

Cavity Block

Core Block

12. Assigning materials:

Materials should be added to the solid bodies to accommodate the correct shrink rate of the plastic part.

Right-click the **Cavity Block** and select the material **Plain Carbon Steel** from the drop-down list (arrow).

Assign the <u>same material</u> to the **Core Block** (Plain Carbon Steel).

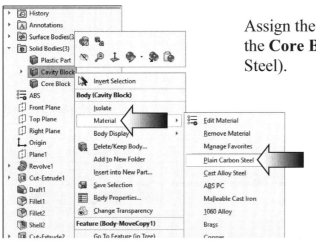

The Plastic Part already has the **ABS** material assigned to it.

13. Changing the transparency:

To further enhance the visibility of the design, we will make the Cavity Block transparent. The default transparency amount is preset to 75%.

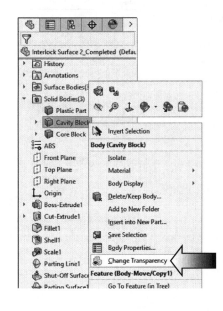

Right-click the Cavity Block and select **Change Transparency** (arrow).

To see the change, click anywhere in the graphics area so that the cavity block is no longer selected.

The Cavity Block is 75% transparent.

To change the transparency amount, select: The Cavity Block, Appearances, Body, Advanced, Illumination, Transparent Amount and drag the slider.

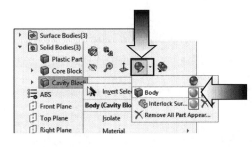

14. Saving your work:

Select **Files, Save As**.

Enter: **Interlock Surface 2_Completed.sldprt** for the file name.

Click **Save**.

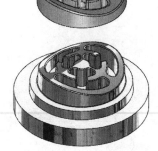

Close all documents.

Exercise – Interlock Surface 3

The interlock surfaces seal the mold properly to prevent leakage. They are also used to maintain the alignment between the mold blocks and keep them from shifting during the molding process.

The interlock surfaces run along the perimeter of the parting surfaces and usually have a draft angle of about 3 to 5 degrees.

This exercise will review the basic steps to prepare a part for a tooling split along with the interlock surfaces.

1. Opening a part document:

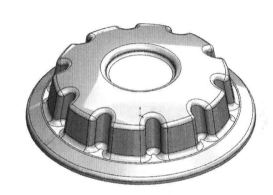

Select **File, Open**.

Open a part document named:
Interlock Surface_Exe.sldprt.

The material **ABS** has already been assigned to this part.

2. Applying the scale factor:

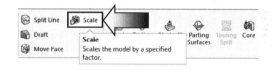

Change to the **Mold Tools** tab and click **Scale** (arrow).

For Scale About, use the default **Centroid** option.

Enable the **Uniform Scaling** checkbox.

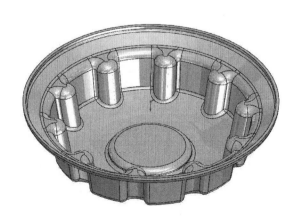

For Scale Factor, enter **1.02** (2% larger).

Click **OK**.

3. Creating the parting lines:

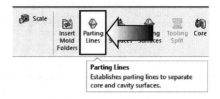

Select the **Parting Lines** command (arrow).

Parting Lines
Establishes parting lines to separate core and cavity surfaces.

For Direction of Pull, select the **Top** plane from the FeatureManager tree.

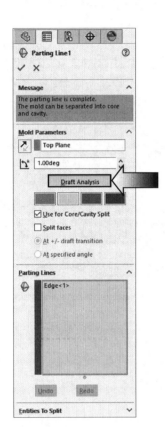

For Draft Angle, enter **1.00deg**.

Click **Draft Analysis** (arrow).

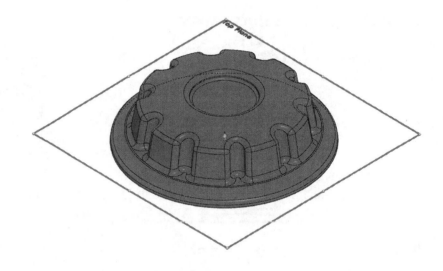

The **bottom edge** of the part is selected automatically. It will be used as the parting line for this part.

Keep all other parameters at their defaults.

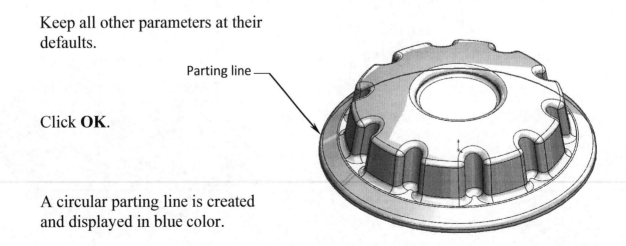

Parting line

Click **OK**.

A circular parting line is created and displayed in blue color.

4. Creating the parting surfaces:

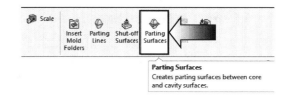

Click **Parting Surfaces** (arrow).

For Mold Parameters, select **Perpendicular to Pull** (arrow).

Parting Line1 is selected automatically.

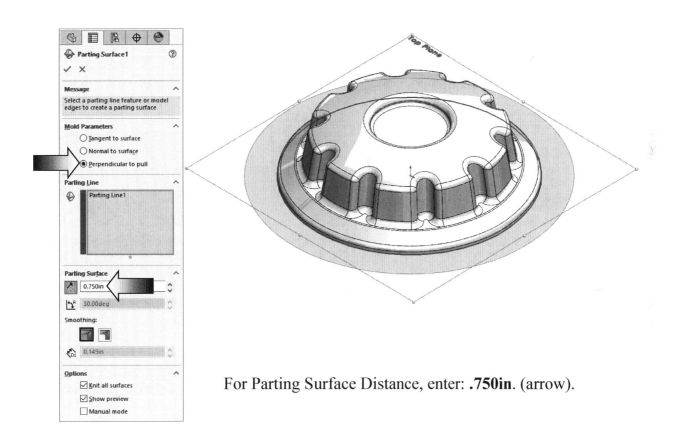

For Parting Surface Distance, enter: **.750in**. (arrow).

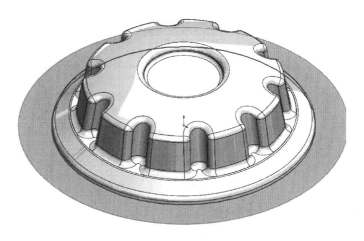

For Smoothing, use the default **Sharp** option.

Enable the **Knit All Surfaces** and **Show Preview** options.

Click **OK**.

5. Adding a new plane:

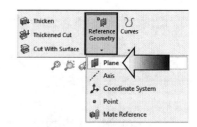

Change to the **Surfaces** tab and select: **Reference Geometry, Plane**.

For First Reference, select the **Top** plane from the FeatureManager tree.

Click the **Offset Distance** option and enter **1.00in**. (arrow).

Click the **Flip Offset** checkbox to place the new plane below the Top plane.

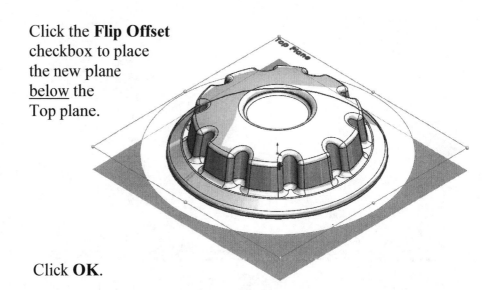

Click **OK**.

6. Making the sketch of the mold block:

Select the new Plane1 and open a **new sketch**.

Sketch a **Circle**, centered on the origin.

Add a diameter dimension to fully define the sketch.

Exit the sketch.

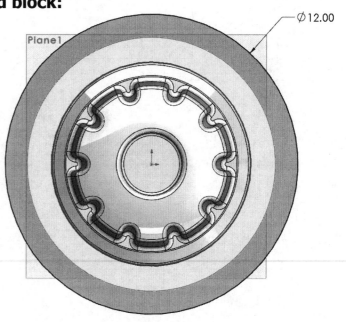

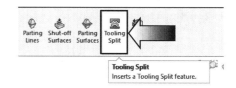

7. Inserting a tooling split feature:

Switch back to the **Mold Tools** tab.

Select the sketch of the **Circle** and click **Tooling Split** (arrow).

For Block Size, enter the following:

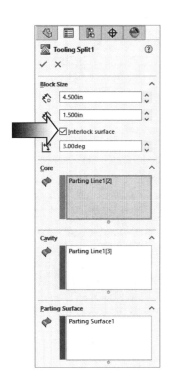

     * **Upper Block = 4.500in**.

     * **Lower Block = 1.500in**.

Enable the **Interlock-Surface** checkbox.

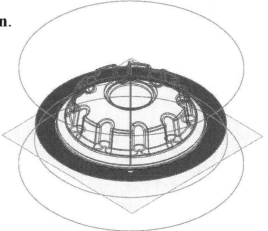

For Draft Angle, enter **3.00deg**.

The surfaces for the Core, Cavity, and Parting Surfaces are populated automatically from the Surface Bodies folder.

Click **OK**.

The 2 mold blocks are created.

Separate the blocks by using the **Move/Copy Bodies** command.

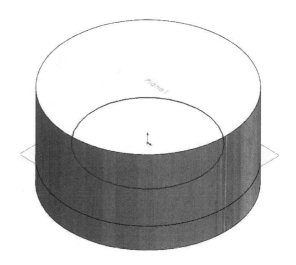

Use a distance of **11 inches** along the **Delta Y** direction for the upper block.

Use a distance of **-9 inches** for the lower block.

8. Assigning materials:

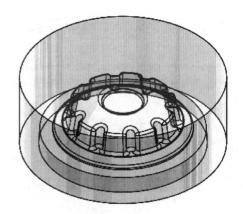

Expand the **Solid Bodies** folder and <u>rename</u> the 3 solid bodies as follow:

> * **Plastic Part**
>
> * **Cavity Block**
>
> * **Core Block**

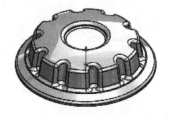

Right-click the **Cavity Block**, select: **Material, Plain Carbon Steel**.

Assign the <u>same material</u> to the **Core Block**.

Right-click the **Cavity Block** and select: **Change Transparency**.

Optional:

Suppress the 2 move/copy steps to collapse the 3 bodies and change the Upper Block and the Lower Block to Transparent, as shown below.

9. Saving your work:

Select **File, Save As**.

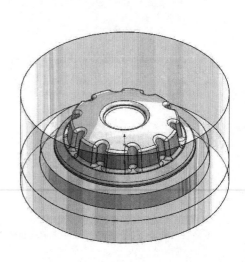

Enter: **Interlock Surface_Exe_Completed** for the file name.

Click **Save**.

Close all documents.

CHAPTER 6

Non-Planar Parting Lines

Using an engineered part, the mold tools are used to create the core and cavity mold. All surfaces from both sides of the part's parting line are copied and knitted into solid blocks to make two mold inserts.

This lesson will teach us one of the easier methods to create a mold for a part with non-planar parting lines.

1. Opening a part document:

Select **File, Open**.

Open a part document named: **Non-Planar Parting Lines.sldprt**

The material **ABS** has already been assigned to this part.

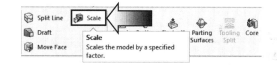

2. Applying the scale factor:

The scale factor provided in the lessons are exclusively for learning purposes. Refer to the material specifications from the manufacturers for more details.

Change to the **Mold Tools** tab and click **Scale**.

For Scale About, select the **Centroid** option.

Enable the **Uniform Scaling** checkbox.

For Scale Factor, enter **1.02** (2% larger).

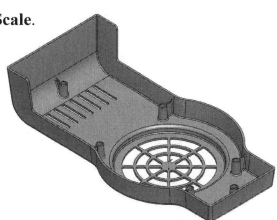

Click **OK**. The part is scaled to 2% larger than its original size.

3. Creating the parting lines:

The parting lines in the part are non-planar due to the step feature on the left side. An interlock surface will be added later on to capture the details of the step.

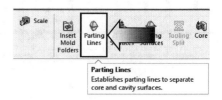

Select the **Parting Lines** command (arrow) on the **Mold Tools** tab.

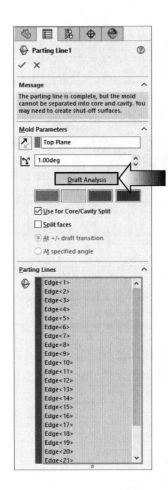

For Direction of Pull, select the **Top** plane.

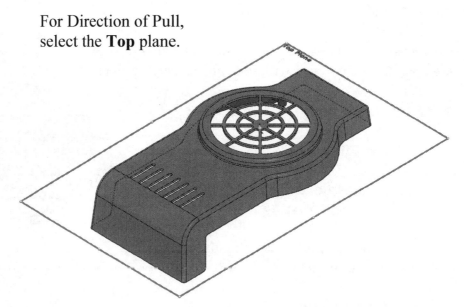

For Draft Angle, enter **1.00deg**.

Click the **Draft Analysis** button (arrow).

The edges along the bottom perimeter of the part are selected automatically.

Click **OK**.

A non-planar parting line is created.

4. Creating the shut-off surfaces:

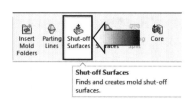

The shut-off surface tool closes up all through-holes by creating a surface patch along the parting line created in the previous step, to define a loop.

Click **Shut-Off Surfaces** (arrow).

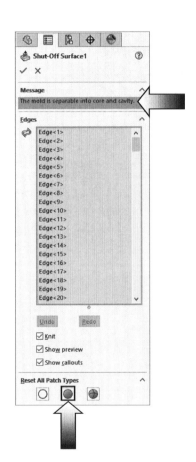

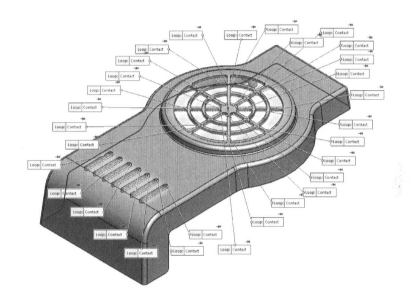

All through-holes are patched up automatically.
Only one Shut-off Surface feature is allowed in each part.
Select **Contact** for fill type. A Fill Type must be assigned to every through hole.

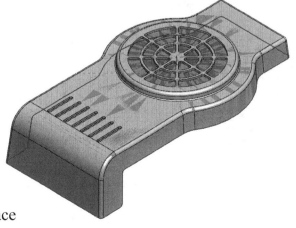

Enable the **Knit**, **Show Preview**, and **Show Callout** checkboxes.

Click **OK**.

All through-holes are closed off.
SOLIDWORKS populates Cavity Surface Bodies and Core Surface Bodies into the Surface Bodies folder with the appropriate surfaces.

5. Creating the parting surfaces:

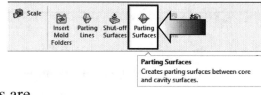

The parting surface split the cavity mold
from the core. It is created by extending
the edges of the model, where the parting lines are,
perpendicular to the pull direction.

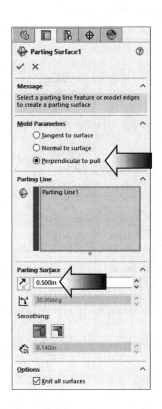

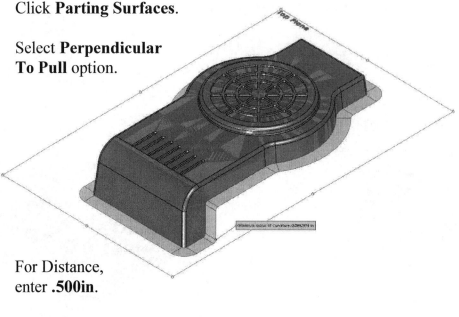

Click **Parting Surfaces**.

Select **Perpendicular
To Pull** option.

For Distance,
enter **.500in**.

Leave other parameters at their defaults.

Click **OK**.

6. Adding a new plane:

Click **Plane** (or select **Insert, Reference Geometry, Plane**).

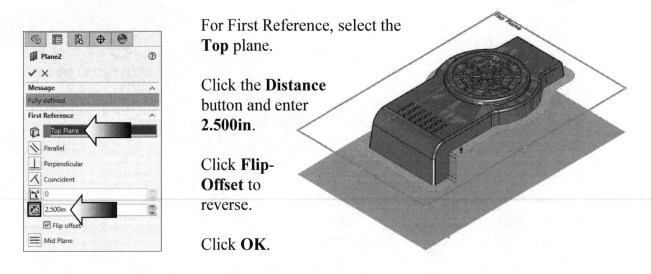

For First Reference, select the
Top plane.

Click the **Distance**
button and enter
2.500in.

Click **Flip-
Offset** to
reverse.

Click **OK**.

7. Sketching the mold block profile:

Select the <u>new plane</u> and open a **new sketch**.

Sketch a **Corner Rectangle** around the part.

Add the dimensions shown in the image to fully define the sketch.

<u>Exit</u> the sketch.

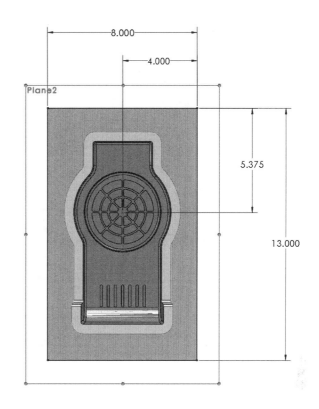

8. Splitting the mold blocks:

Switch to the **Mold Tools** tab.

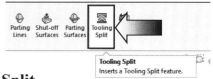

Select the sketch of the **Rectangle** and click **Tooling Split**.

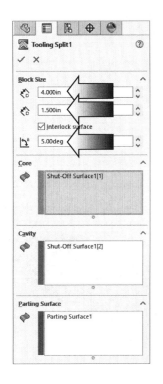

For Block Size, enter:

    Upper Block = 4.000in
    Lower Block = 1.500in

Enable the **Interlock-Surface** Checkbox.

For Draft, enter **5 deg**.

Click **OK**.

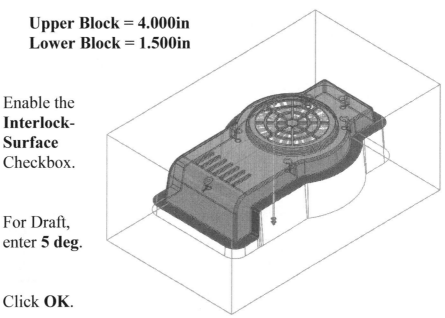

The 2 mold blocks are created and currently shown in the closed position.

We will separate the 2 mold blocks after hiding some of the reference surfaces and the parting lines.

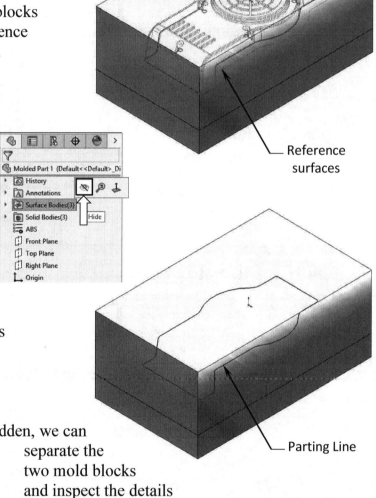

Reference surfaces

9. Hiding the references:

Select the **Surface Bodies** folder and click **Hide** (arrow).

To hide the parting line, select: **View, Hide/Show, Parting Lines**, or right-click the parting line in the graphics area and select Hide.

Parting Line

Now that all references are hidden, we can separate the two mold blocks and inspect the details of the core and cavity.

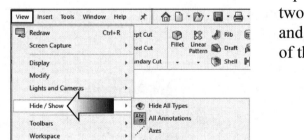

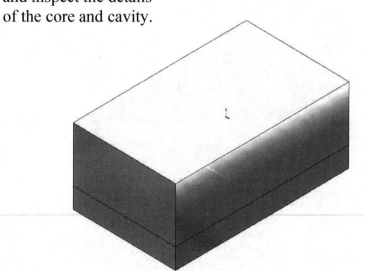

10. Separating the mold blocks:

Click **Move/Copy Bodies** (or select: **Insert, Features, Move/Copy**).

For Bodies to Move/Copy, select the **upper block** (the Cavity).

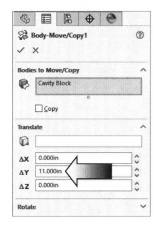

Click in the <u>Delta Y</u> box and enter **11.000in**.

Click **OK**.

The upper block moves upward 11 inches.

Click **Move/Copy Bodies** again.

For Bodies to Move, select the **lower block** (the Core).

Click in the <u>Delta Y</u> field and enter **-9.000in**.

Click **OK**.

The lower block moves downward 9 inches.

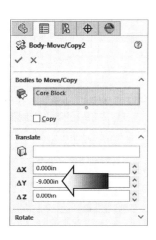

The 2 move features appear on the Feature-Manager tree and can be suppressed to collapse the mold.

11. Changing the transparency:

For clarity, we will rename the solid bodies and change the Cavity block to transparent, that way its details can be seen using the default isometric view.

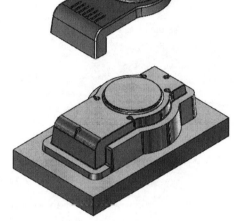

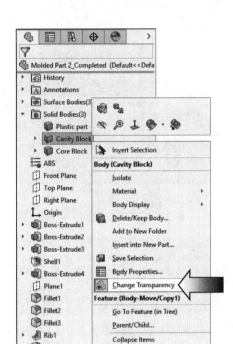

Right-click the Cavity Block and select: **Change Transparency**.

Leave the Plastic Part and the Core Block as solid shaded.

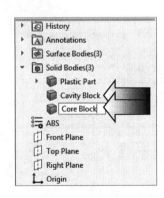

The engineered part is renamed to **Plastic Part**.

Rename the upper block to **Cavity Block**.

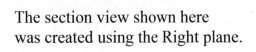

Rename the lower block to **Core Block**.

Section Views are occasionally used to verify the wall thickness or the parting lines of the part.

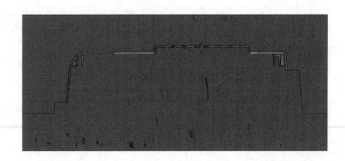

The section view shown here was created using the Right plane.

12. Changing the materials:

The Plastic part should already have the **ABS** material assigned to it.

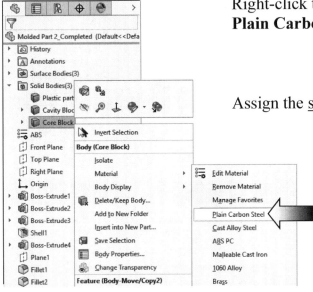

Right-click the **Cavity Block**, click **Material**, and select **Plain Carbon Steel**.

Assign the <u>same material </u> to the **Core Block**.

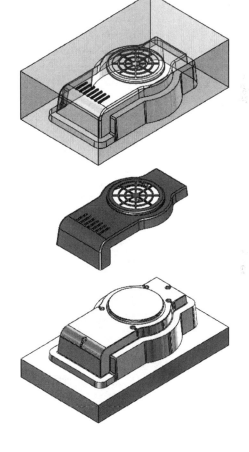

There are additional steps to complete the mold such as runners and gates, cooling lines and other parts to position the inserts inside a mold base. These steps are usually done by a tool designer or mold maker. We will not discuss those steps here in the lessons.

13. Saving your work:

Select **File, Save As**.

Enter: **Non-Planar Parting Lines_Completed.sldprt** for the file name.

Click **Save**.

Close all documents.

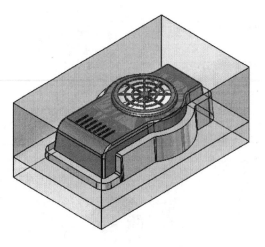

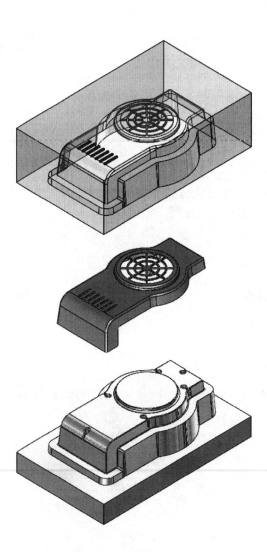

<u>Exercise – Non-Planar Parting Lines</u>

A lot of times, the parting surfaces tool may not be able to create the surfaces needed due to the complex non-planar parting lines in a model. In this exercise, we will take a look at other techniques where the shut-off surfaces, parting surfaces, and interlocks are created manually.

1. Opening a part document:

Select **File, Open**.

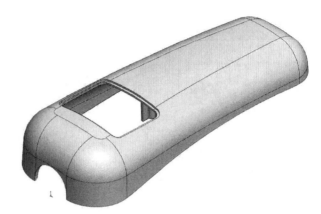

Open a part document named:
**Non-Planar parting Lines_Exe.
sldprt**.

The material **ABS** has already been assigned
to the part.

2. Applying scale:

Change to the **Mold Tools** tab.

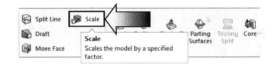

Click **Scale** (arrow).

For Scale About, select
the **Centroid** option.

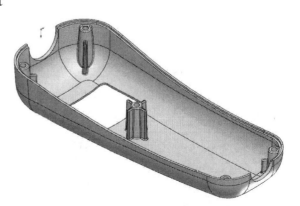

Enable the **Uniform-
Scaling** checkbox.

For Scale Factor, enter **1.02** (2% larger).

Click **OK**.

3. Creating the parting lines:

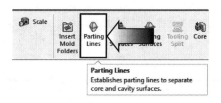

Click **Parting Lines** (arrow).

For Direction of Pull, select the **Top** plane from the FeatureManager tree.

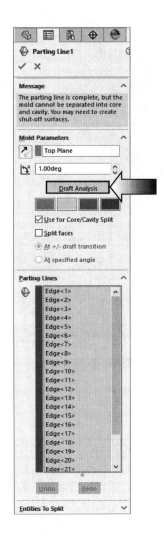

For Draft Angle, enter **1.00deg**.

Click **Draft Analysis**.

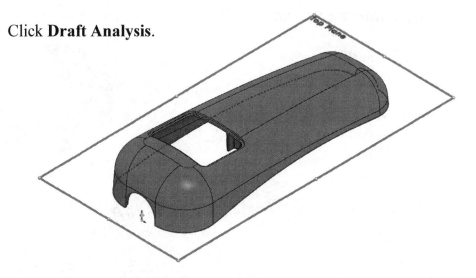

The **bottom edges** of the model are selected automatically and a parting line is created from these edges.

The parting line lies between the Green surfaces (used to make the cavity block) and the Red surfaces (used to make the Core block).

Click **OK**.

A parting line is created. It runs along the bottom edges of the part and is displayed in blue color.

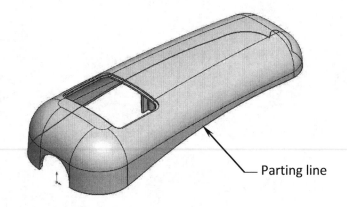

Parting line

4. Creating the parting surfaces:

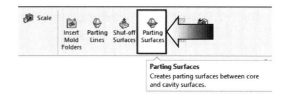

Select the **Parting Surfaces** command (arrow).

Parting Surfaces
Creates parting surfaces between core and cavity surfaces.

For Mold Parameters, select **Perpendicular to Pull** (arrow).

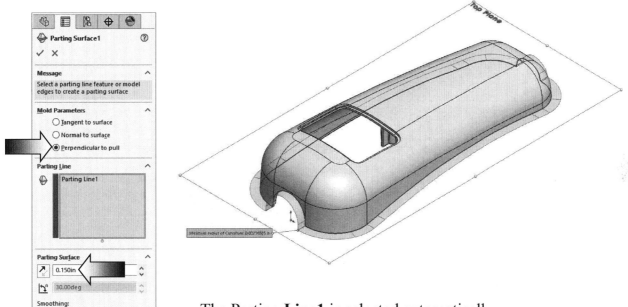

The Parting **Line1** is selected automatically.

For Parting Surface Distance, enter **.150in**. (arrow).

Keep all other parameters at their default settings.

Click **OK**.

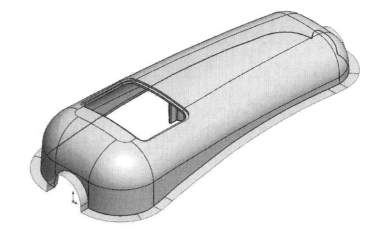

A parting surface is created. It is protruded outward, perpendicular to the parting line.

5. Adding the ruled surfaces:

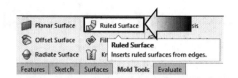

The ruled surfaces will be used as the interlock surfaces during the tooling split.

Click **Ruled Surface** (arrow).

For Type, select **Taper to Vector** (arrow).

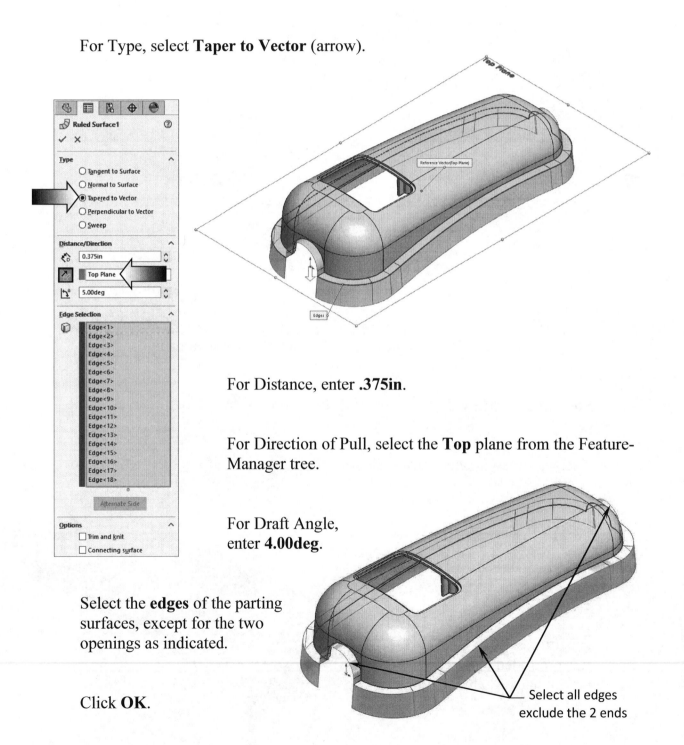

For Distance, enter **.375in**.

For Direction of Pull, select the **Top** plane from the Feature-Manager tree.

For Draft Angle, enter **4.00deg**.

Select the **edges** of the parting surfaces, except for the two openings as indicated.

Click **OK**.

Select all edges exclude the 2 ends

6. Closing off the left end:

It will take a few steps such as loft, extend, and trim to patch up the openings. The first step is to create a lofted surface between two edges.

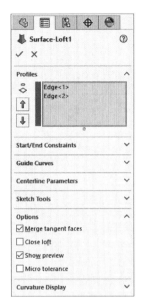

Change to the **Surfaces** tab and click **Lofted Surface**.

For Loft Profiles, select the **2 edges** as noted.

Keep all other parameters at their defaults.

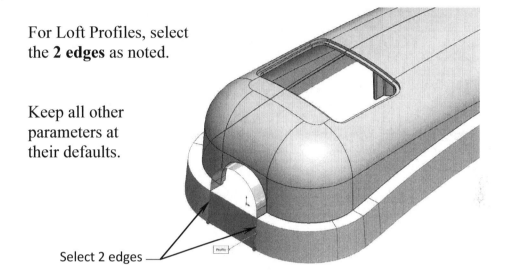

Select 2 edges

Click **OK**.

7. Extending the lofted surface:

The ruled surfaces have a 5° draft. We will need to extend and trim the lofted surface to maintain that angle.

Click **Extend Surface**.

Select the edge as indicated and enter **.600in**. for distance (arrow).

For Type, click **Same Surface**.

Click **OK**.

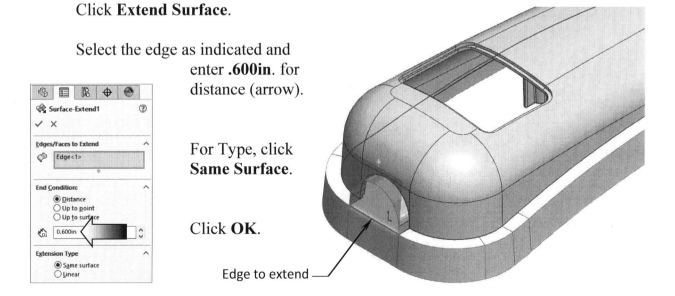

Edge to extend

8. Trimming the left end:

The lofted surface maintains its 5° draft. We will now trim it to the final shape.

Click **Trim Surface**.

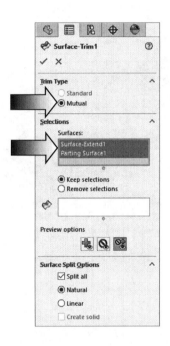

For Trim Type, select the **Mutual** option.

For Selections, select the **2 faces** as noted.

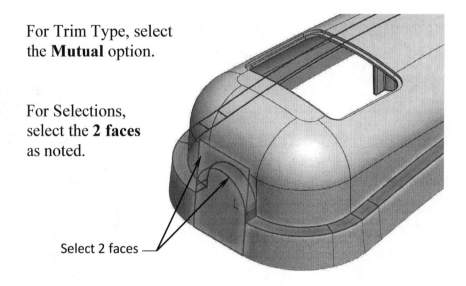

Select 2 faces

Click the **Remove Selection** button (arrow).

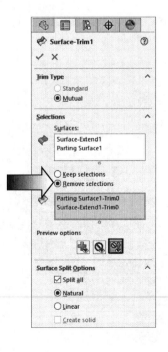

For Surfaces to Remove, select the **upper portion** of the lofted surface and the small, **outer curved portion** as indicated.

Under Preview Options, change to different options such as Show Included, Show Excluded, or **Show Both**, to preview the trim.

Click **OK**.

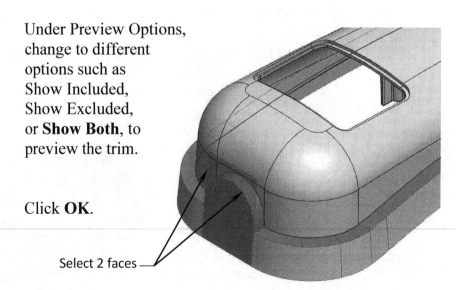

Select 2 faces

9. Closing off the right end:

The right end of the model also needs to be patched up the same way as the left end.

Switch back to the **Surfaces** tab and click **Lofted Surface**.

For Loft Profiles, select the **2 edges** as indicated.

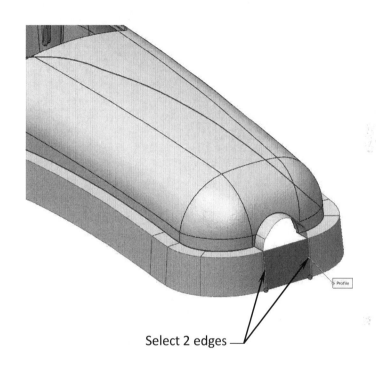

Select 2 edges —

Leave all other parameters at their defaults.

Click **OK**.

It would be quicker to patch up the opening using the Filled Surface tool, but we will not have the 5° draft angle to match up with the ruled surfaces. Let us continue with using the same method as we did in the previous step.

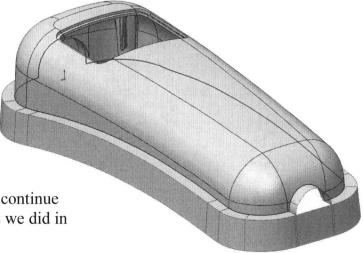

10. Extending the lofted and ruled surfaces:

Select the **Extend Surface** tool.

For Edges/Faces to Extend, select the **3 edges** as noted.

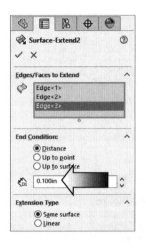

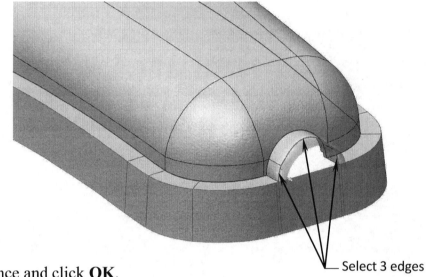

Select 3 edges

Enter **.100in**. for distance and click **OK**.

Click **Extend Surface** again.

For Edges/Faces to Extend, select the **edge** shown below.

Enter **.300in**.
for distance.

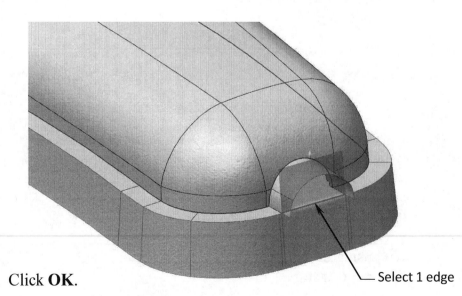

Select 1 edge

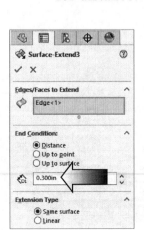

Click **OK**.

11. Trimming the right end:

Select the **Trim Surface** command.

For Trim Type, select the **Mutual** option (arrow).

For Selections, select the **2 surfaces** as indicated.

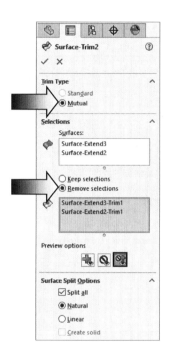

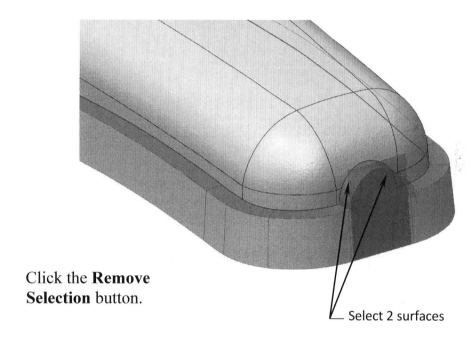

Select 2 surfaces

Click the **Remove Selection** button.

For surfaces to remove, select the **upper portion** of the lofted surface and the **outer curved portion** as indicated in the image above.

For Preview Options, select **Display Both Included and Excluded surfaces**.

Leave other settings at their defaults.

Click **OK**.

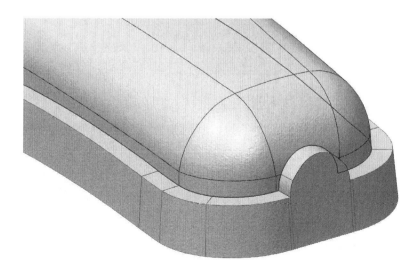

12. Knitting all surfaces:

All of the surfaces needed to make up the interlock surface are created; they should be knitted into a single surface.

Select the **Knit Surface** command.

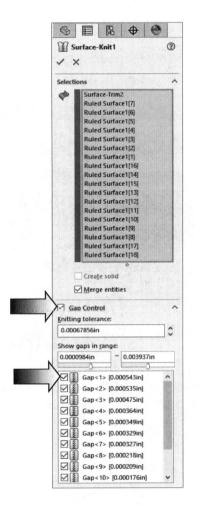

Expand the Surface Bodies folder and select **all surfaces** inside of the folder, or select them directly from the graphics area.

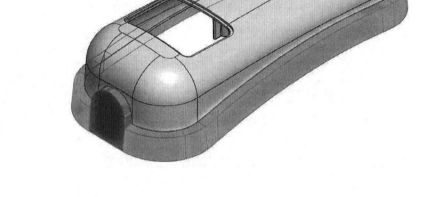

Enable the **Merge Entities** checkbox.

Click the **Gap Control** checkbox and enable all of its checkboxes to allow the gaps to be healed automatically.

Click **OK**.

All surfaces are knitted into a single surface. This surface will be used as the new interlock surface when the tooling split is inserted.

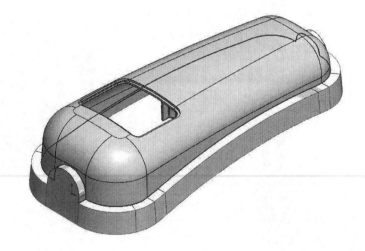

13. Creating the shut off surfaces:

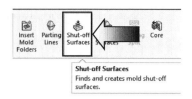

The window on top of the part must be closed off.
The shut-off surfaces tool is used for this.

Change to the **Mold Tools** tab and click **Shut-Off Surfaces** (arrow).

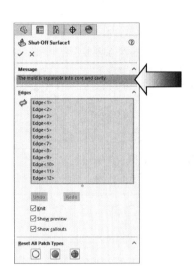

All edges of the opening
are selected and a
new surface is
created to
close it off.

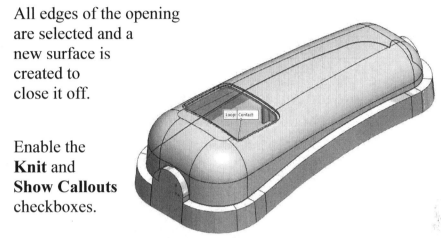

Enable the
Knit and
Show Callouts
checkboxes.

Click **OK**.

14. Adding a new plane:

Switch back to the **Surfaces** tab and select **Reference Geometry, Plane**.
This plane is used to sketch the profile of the mold block and also to set the height
of the interlock surface.

For First Reference, select the **Top** plane.

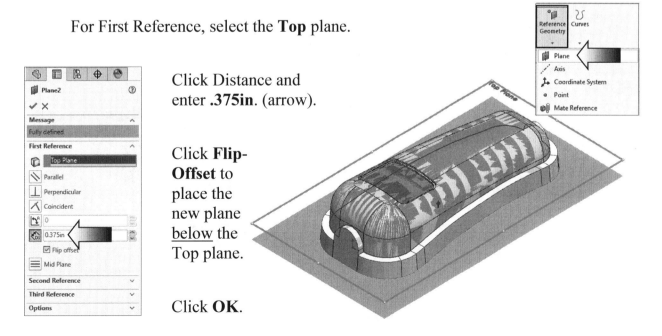

Click Distance and
enter **.375in**. (arrow).

Click **Flip-Offset** to
place the
new plane
below the
Top plane.

Click **OK**.

15. Inserting a tooling split feature:

The Tooling Split tool uses a sketch to create the core and cavity blocks for a mold.

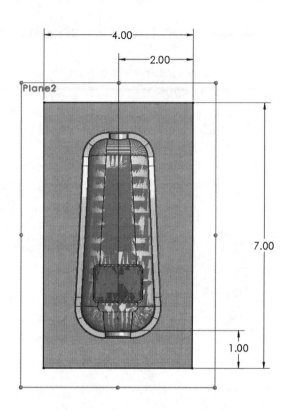

Select the new <u>Plane2</u> and open a **new sketch**.

Sketch a **Corner Rectangle** and add the dimensions shown to fully define it.

<u>Exit</u> the sketch.

Under Block Size, enter the following:

    * Upper Block = 2.00in.
    * Lower Block = 1.00in.

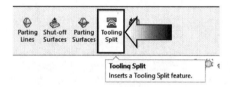

Enable the **Interlock Surface** checkbox and enter **5.00deg** for draft angle.

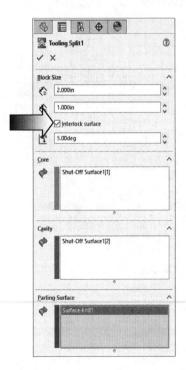

The surfaces for the Core, Cavity, and Parting Surfaces are populated automatically.

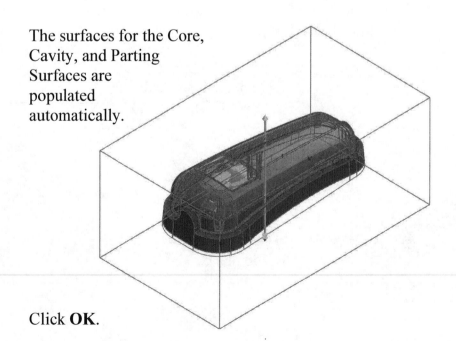

Click **OK**.

16. Separating the mold blocks:

Change to the **Features** tab and click **Move/Copy Bodies**.
(or select: **Insert, Features, Move/Copy**).

For Bodies to Move/Copy, select the **upper block** (the Cavity).

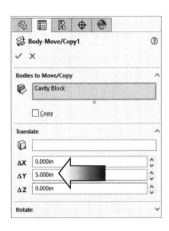

Click in the <u>Delta Y</u> field and enter **5.00in**. (arrow).

Click **OK**.

The upper block moves upward 5 inches.

Click **Move/Copy Bodies** again.

For Bodies to Move/Copy, select the **lower block** (the Core).

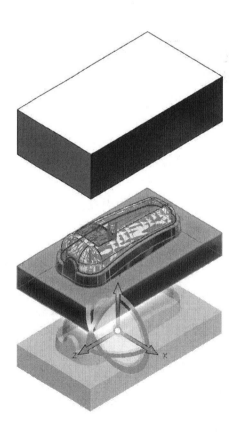

Click in the <u>Delta Y</u> field and enter **-3.500in**.

Click **OK**.

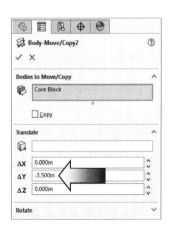

The lower block moves downward 3.5 inches.

Hide all reference surfaces. For clarity, also hide the parting lines.

17. Changing materials:

Expand the Solid Bodies folder and <u>rename</u> the 3 solid bodies to:

Plastic Part

Cavity Block

Core Block

Right-click the **Cavity Block**, select: **Material, Plain Carbon Steel**.

Assign the <u>same material</u> to the **Core Block**.

The Plastic Part has the **ABS** material assigned to it already.

Right-click the Cavity Block and select: **Change Transparency**.

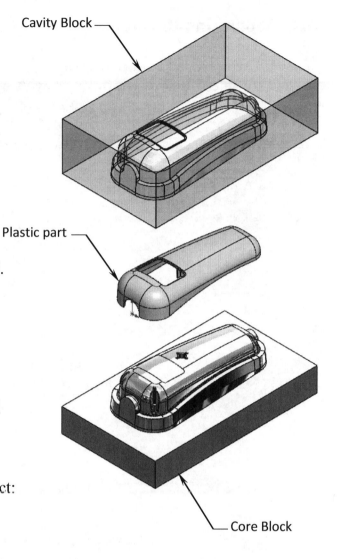

Cavity Block

Plastic part

Core Block

18. Saving your work:

Select **File, Save As**.

Enter **Non_Planar Parting Lines_Exe_ Completed.sldprt** for the file name.

Click **Save**.

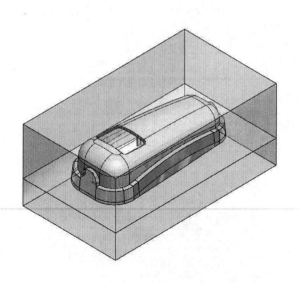

Close all documents.

CHAPTER 7

Manual Parting Lines

When a part has non-planar parting lines,
the interlock surface is usually created manually and knitted to the parting surface of the part. Other surface modeling tools such as ruled surface, lofted surface, planar surface, filled surface and trim surface are also used to complete the interlock surface.

This lesson will teach us one of the methods to create the non-planar parting lines and parting surfaces manually.

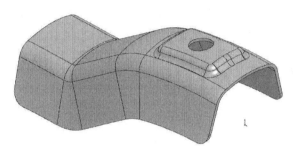

1. Opening a part document:

Select **File, Open**.

Open a part document named:
Manual Parting Lines.sldprt.

2. Applying the scale factor:

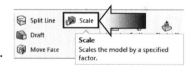

The material **ABS** has already been assigned to the part.

Change to the **Mold Tools** tab and select the **Scale** command.

For Scale About, select the **Centroid** option.

Enable the **Uniform Scaling** checkbox.

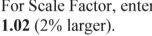

For Scale Factor, enter **1.02** (2% larger).

Click **OK**.

The part is scaled 2% larger to accommodate the shrink rate of the selected material.

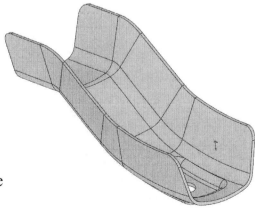

3. Creating the parting lines:

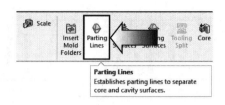

Parting Lines
Establishes parting lines to separate core and cavity surfaces.

When creating the parting lines, a plane (or a planar face) is used to define the direction that the cavity body is pulled to split the core and cavity.

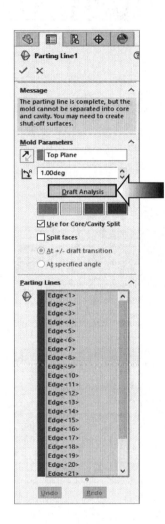

Switch back to the **Mold Tools** tab and click **Parting Lines**.

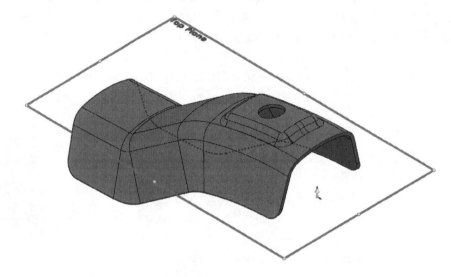

For Direction of Pull, select the **Top** plane from the FeatureManager tree.

For Draft Angle, enter **1.000deg**.

Click **Draft Analysis** (arrow).

Click **OK**.

The Parting Line for the part is created and displayed in Blue color.

Since the parting line is non-planar, additional surfaces will need to be created and knitted to the parting surface.

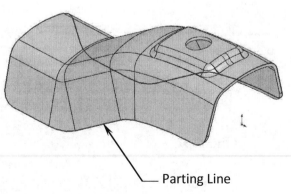

Parting Line

4. Creating the shut-off surfaces:

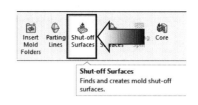

The shut-off surface tool closes up all through-holes by creating a surface patch along the parting line that was created in the last step to define a loop.

Click **Shut-Off Surfaces** (arrow).

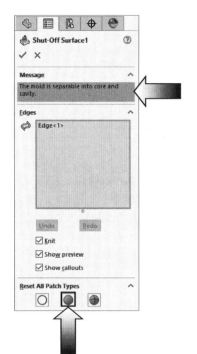

The **circular edge** of the hole is selected automatically.

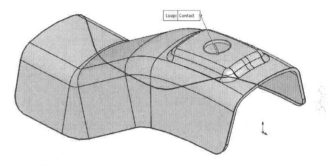

A green-color message indicates "the mold is separable into core and cavity".

Enable the checkboxes for **Knit, Show Preview**, and **Show-Callouts**.

Under Patch Types, select the **All Contact** option (arrow).

Click **OK**.

Expand the Surface Bodies folder and click on either the Cavity Surface Bodies or the Core Surface Bodies to see the Red and Green surfaces.

The Red color represents the Cavity, and the Green color represents the Core.

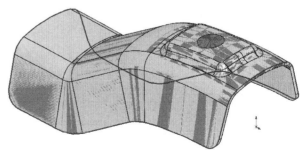

5. Creating the parting surfaces:

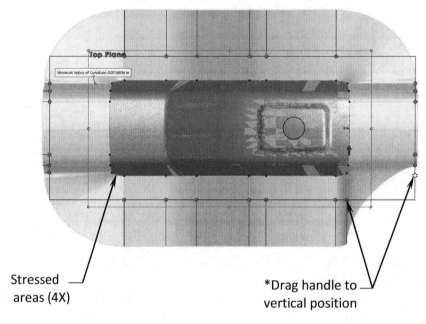

The parting surfaces are used to split the mold cavity from the core. Because of the complex parting lines in this model, several surfaces will need to be built manually to make up the custom parting surfaces.

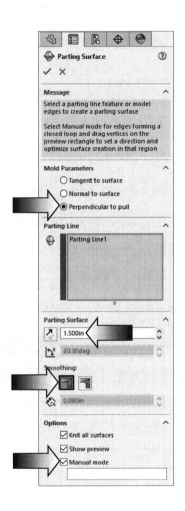

Click **Parting Surfaces** (arrow).

Select the **Perpendicular to Pull** option (arrow).

Stressed areas (4X)

*Drag handle to vertical position

For Parting Surface Distance, enter **1.5000in**.

For Smoothing, select **Sharp**.

The 4 corners are deformed which can cause undesired results.

*Enable the **Manual Mode** checkbox and <u>drag</u> the <u>handles</u> as noted to smooth out the 4 corners.

Click **OK**.

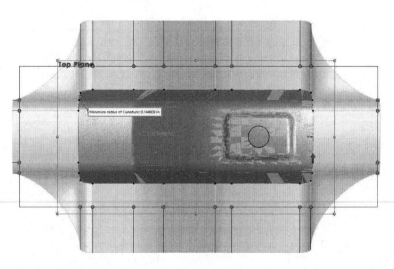

6. Deleting the surfaces:

The 4 corner surfaces still not tangent with the parting surfaces. They need to be deleted and then recreated manually.

Switch to the Surfaces toolbar and click **Delete Face**.

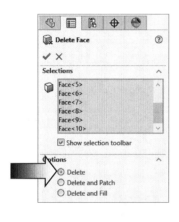

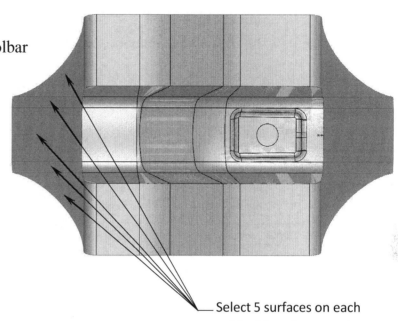

Select 5 surfaces on each side, 10 surfaces total

Click the **Delete** option.

Select the 10 surfaces as indicated.

Click **OK**.

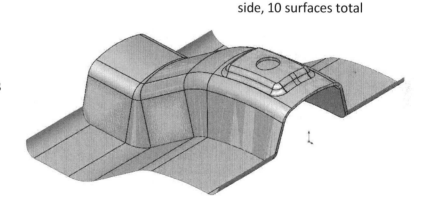

The selected surfaces are deleted.

The front, back, and 4 corners will be reconstructed in the next few steps.

(The completed parting surfaces is shown here for reference.)

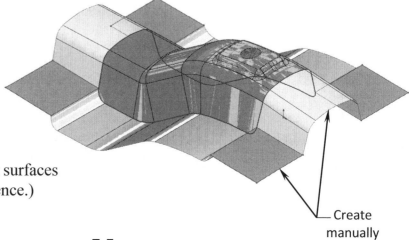

Create manually

7. Creating the ruled surfaces:

The Ruled Surface tool creates surfaces that protrude out in a specified direction from a group of selected edges. *(Hide the parting Lines if needed when selecting the edges.)*

Select the **Rule Surface** command (arrow) from the Mold Tools tab.

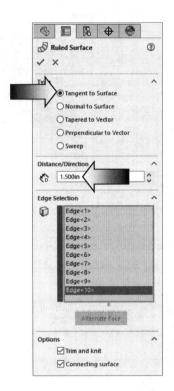

For Type, select **Tangent To Surface** (arrow).

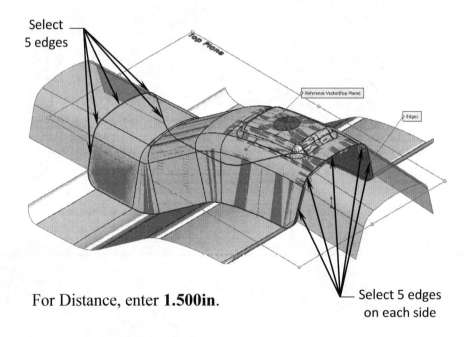

Select 5 edges

Select 5 edges on each side

For Distance, enter **1.500in**.

For Edge Selection, select **5 edges on each side** of the model (10 edges total).

All edges should extend outward. To reverse any of the edges, select them from the Edge Selection window and click: **Alternate Face**.

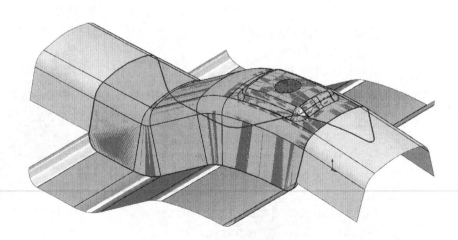

Click **OK**.

8. Trimming with a sketch:

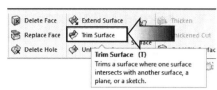

Open a **new sketch** on the <u>Front</u> plane.

Sketch **2 lines** on both sides of the model and add the
dimensions / relations shown. These lines will be used to trim the bottom to a flat.

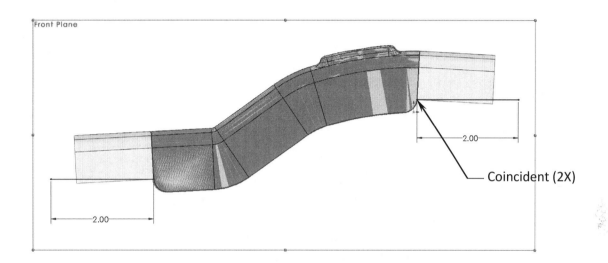

Switch to the **Surfaces** tab and click **Trim Surface**. (or click **Insert, Surface, Trim**).

For Trim Type, click **Standard**.

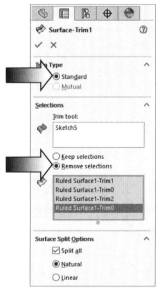

For Trim Tool, select the sketch of the **2 lines**.

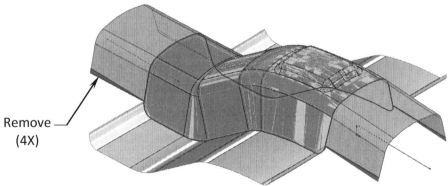

Remove
(4X)

Click the **Remove Selections** option and select the **four triangular portions** of the ruled surfaces, as noted.

Click **OK**.

9. Making a 3D sketch:

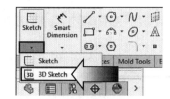

The 4 corners of the ruled surfaces must be closed off. We will use a single 3D sketch to enclose the ends.

Select **3D Sketch** from the Sketch drop down (arrow).

Sketch **2 Lines** at each corner of the ruled surfaces (8 lines total).

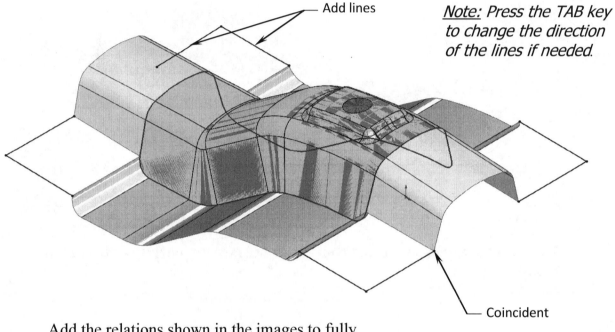

Add lines

Note: Press the TAB key to change the direction of the lines if needed.

Coincident

Add the relations shown in the images to fully define the sketch (skip the relations if the lines are drawn perfectly straight).

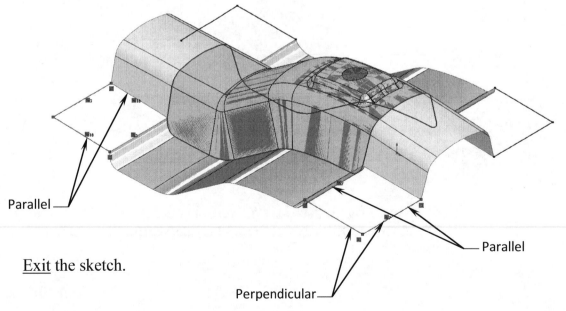

Parallel

Parallel

Perpendicular

Exit the sketch.

10. Creating the 1st corner patch:

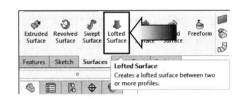

The Loft tool creates a feature by making transitions between the profiles.

Change to the **Surfaces** tab and click **Lofted Surface** (arrow) or select: **Insert, Surface, Loft**.

For Loft Profile, select the **1 line** and **1 edge** as indicated. Click OK on the SelectionManager pop-up dialog box.

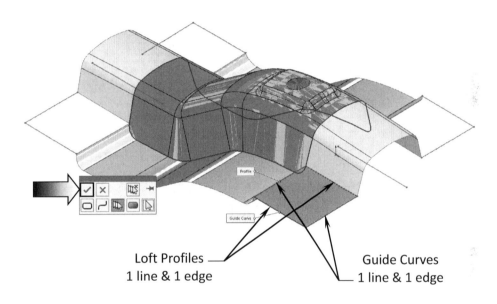

Loft Profiles — 1 line & 1 edge

Guide Curves — 1 line & 1 edge

For Guide Curves, select **1 line** and **1 edge** as noted. Click OK on the SelectionManager pop-up dialog box.

The SelectionManager pop-up dialog box allows you to select one or more entities in a sketch and use as a loft profile or as a guide curve.

Click **OK**.

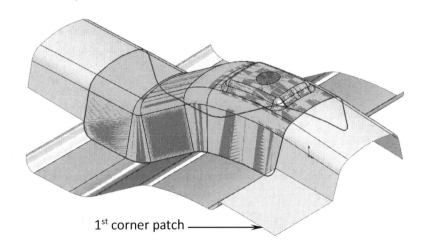

1st corner patch →

11. Patching other corners:

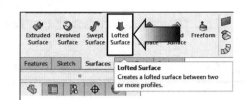

Repeat step number 9 and create a total of
4 lofted surfaces, one at each corner.
(Show the 3D Sketch under the Surfaces-Loft1.)

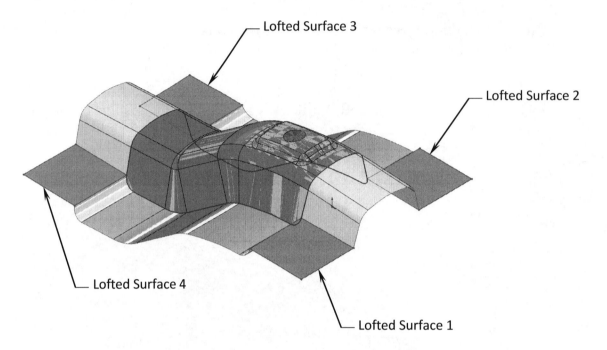

Zoom in on each corner and check the lofted surfaces for gaps.

Press **Control + 7** to change to the **Isometric** view.

Hold the **Shift** key and push the
Up Arrow key <u>twice</u> to change
to the **reverse Isometric**
view.

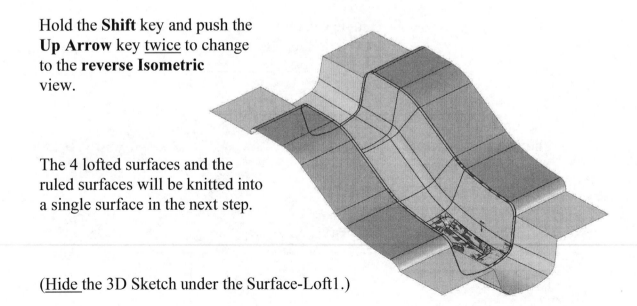

The 4 lofted surfaces and the
ruled surfaces will be knitted into
a single surface in the next step.

(<u>Hide</u> the 3D Sketch under the Surface-Loft1.)

12. Knitting the surfaces:

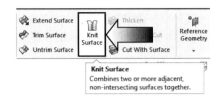

The Knit Surface tool combines two or more surfaces into a single surface.

Click **Knit Surface** (arrow).

For Selection, select **all surfaces** in the graphics area (or expand the Surface Bodies folder (arrow) and select all surfaces from there).

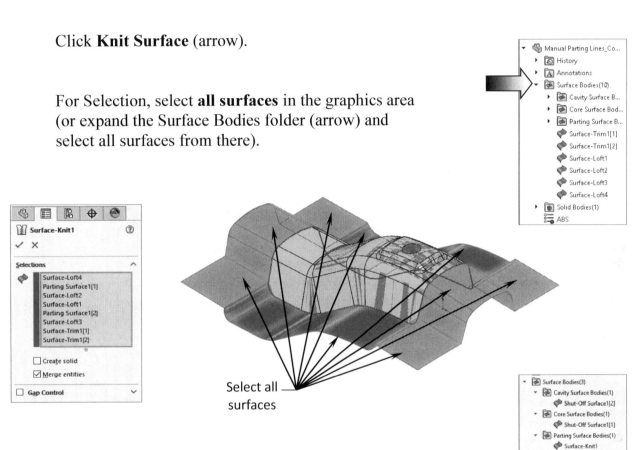

Select all surfaces

Enable the **Merge Entities** checkbox.

If there are no gaps between the surfaces, uncheck the Gap Control checkbox. But if gaps are found among the surfaces, enabling this checkbox will allow the Knit tool to heal them.

Click **OK**.

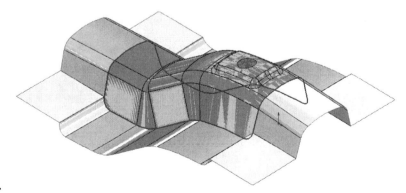

All surfaces are knitted into one. There should be one surface inside each surface bodies folder.

13. Creating a reference plane:

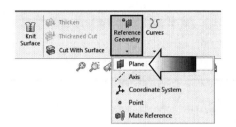

Since the Knit Surface is a non-planar surface, a new plane is needed to make the sketch of the mold block.

Click **Plane** (or select **Insert, Reference Geometry, Plane**).

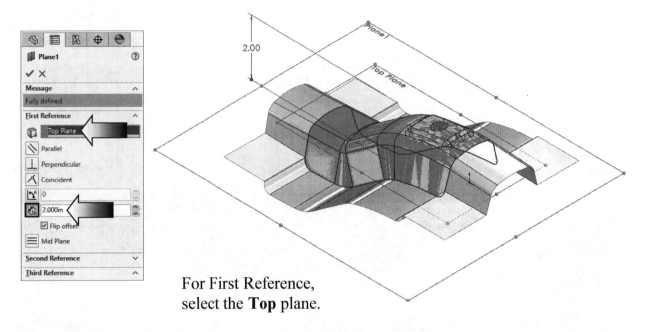

For First Reference, select the **Top** plane.

Click the Distance button and enter **2.000in**.
Place the new plane <u>below</u> the Top plane; click Flip Offset if needed.

Open a **new sketch** on the <u>Plane1</u>.

Sketch a **Corner-Rectangle** and add the dimensions shown in the image.

<u>Exit</u> the sketch.

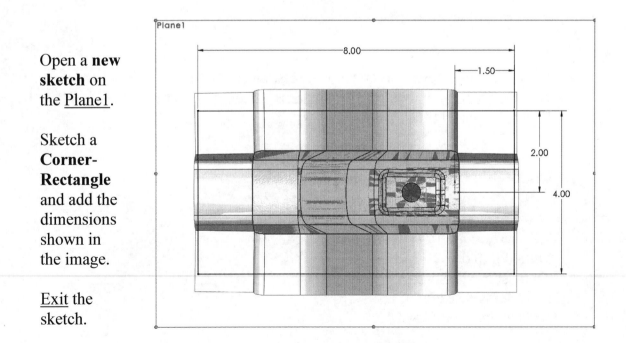

14. Creating a tooling split:

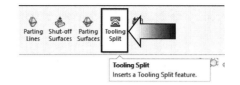

The Tooling Split tool is used to create the core and cavity blocks for a mold.

Change to the **Mold Tools** tab.

Select the sketch of the <u>rectangle</u> and click **Tooling Split**.

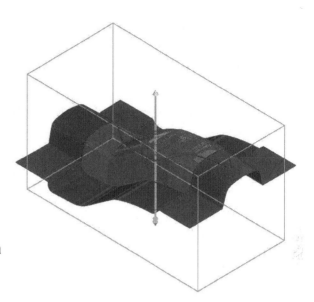

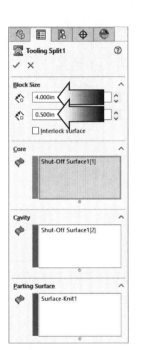

For Block Size, enter:

Upper Block = 4.000in

Lower Block = .500in

The Core, Cavity, and Parting Surfaces are selected automatically. <u>Clear</u> the Interlock Surface checkbox.

Click **OK**.

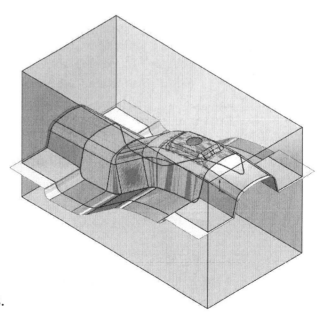

The two mold blocks are created.

The mold blocks were made transparent for clarity only.
Only the Cavity block will be made transparent in the next couple of steps.

15. Separating the mold blocks:

When separating the mold blocks, the Move/Copy command is preferred over the Exploded View command. The Move Steps shows up on the FeatureManager tree, and can be suppressed at anytime to collapse the blocks.

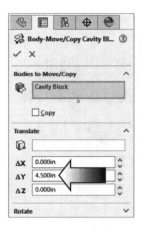

Click **Move/Copy Bodies** (or select **Insert, Features, Move/Copy**).

For Bodies to Move, select the **upper block** (the Cavity).

Click in the <u>Delta Y</u> field and enter **4.500in** (arrow)

Click **OK**. The cavity block moves upward 4.5 inches.

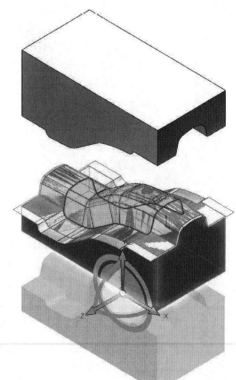

Select the **Move/Copy** command again.

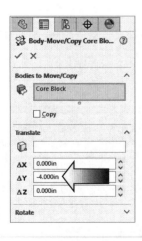

For Bodies to Move, select the **lower block** (the Core).

Click in the Delta Y field and enter **-4.000in** (arrow).

Click **OK**.

The Core block moves downward 4 inches.
The reference surfaces will be put away in the next step.

16. Hiding the references:

For clarity, the reference surfaces and the parting lines should be hidden at this point.

Click the **Surface Bodies** folder and select **Hide**.

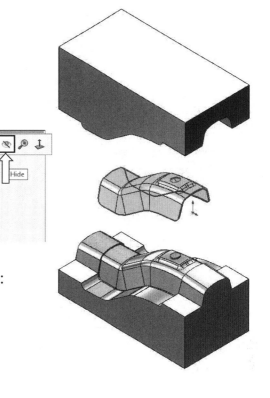

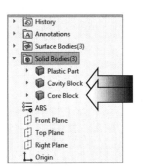

Rename the Solid Bodies to:
> **Plastic part**
> **Core Block**
> **Cavity Block**

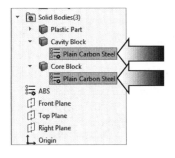

Right-click the **Core Block** and select the material: **Plain Carbon Steel** (arrow).

Assign the same material to the **Cavity Block** (arrow).

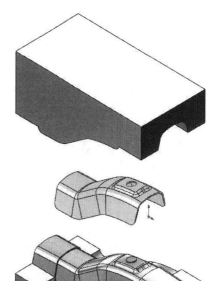

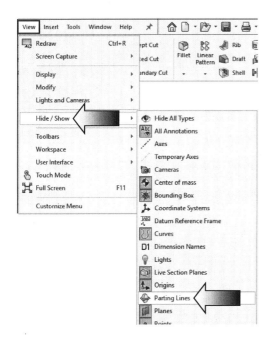

To hide the parting lines, select: **View, Hide/Show, Parting Lines**.

The Cavity Block will be changed to transparent in the next step.

17. Changing to transparency:

The Cavity Block is often made transparent to use in design review sessions, or to take screenshots and use in PowerPoint slides, etc.

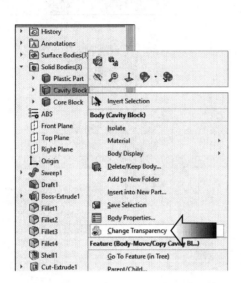

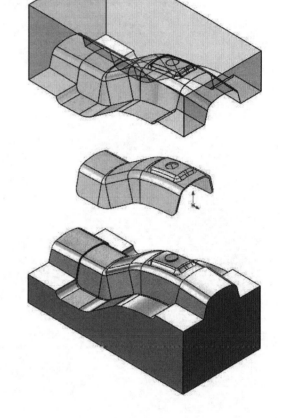

Under the Solid Bodies folder, right-click the Cavity Block and select:
Change Transparency (arrow).

The Cavity Block is changed to transparent. (The default transparency amount is 75%.)

18. Saving your work:

Select **File, Save As**.

Enter: **Manual Parting Lines_Completed.sldprt** for the file name.

Click **Save**.

Close all documents.

CHAPTER 8

Undercuts and Slide Cores

The Undercut Analysis tool is used to find trapped areas in a plastic part that cannot be ejected from the mold. These areas require one or more slide cores. When the main core and cavity are separated, the slide core(s) slides in a direction perpendicular to the motion of the main core and cavity. This allows the part to be ejected from the mold.

This lesson will teach us how a slide core is created to capture the connector ports on the left end of the part.

1. Opening a part document:

Select **File, Open**.

Open a part document named: **Slide Core.sldprt**

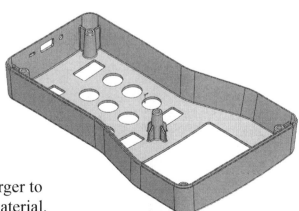

2. Applying the scale factor:

The material **ABS** has already been assigned to this model.

Change to the **Mold Tools** tab and click **Scale**.

For Scale About, select the **Centroid** option from the list.

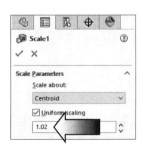

Enable the **Uniform-Scaling** checkbox.

For Scale Factor, enter **1.02** (2% larger).

Click **OK**.

The overall size of the part grows 2% larger to accommodate for the shrinkage of the material.

3. Creating the parting lines:

Select the **Parting Lines** command on the **Mold Tools** tab.

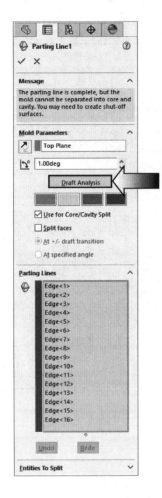

For Direction of pull, select the **Top** plane from the Feature-Manager tree.

For Draft Angle, enter **1.00deg**.

Click **Draft Analysis** (arrow).

The **bottom edges** of the part are selected automatically as the parting lines for this model.

A **yellow color** message appears on top of the tree indicating the parting line is complete, but the mold cannot be separated into core and cavity. Additional steps will need to be added.

Click **OK**.

A planar parting line is created.

4. Creating the shut-off surfaces:

The shut-off surface tool closes up a through hole by creating a surface patch along the edges that forms a continuous loop, or in this case, the parting line of the part.

Click **Shut-Off Surfaces**.

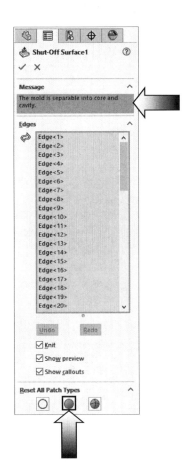

The edges of all through openings in the part are selected and shut off automatically.

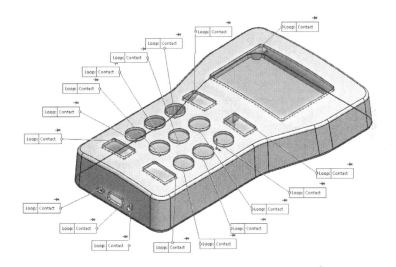

A green-color message appears on the top left of the properties tree indicating the mold is separable into core and cavity.

Enable the **Knit**, **Show Preview**, and Show **Callouts** checkboxes.

For Patch Type, use the default **All-Contact** (arrow).

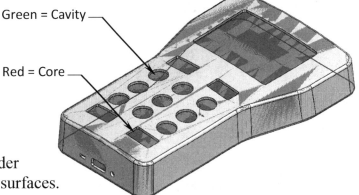

Green = Cavity

Red = Core

Click **OK**.

Expand the **Surface Bodies** folder to view the red and green color surfaces.

5. Creating the parting surfaces:

Parting surfaces tool splits the mold cavity from the core.

Click **Parting Surfaces**.

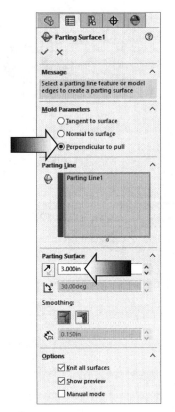

For Mold Parameters, select **Perpendicular to Pull** (arrow).

The **Parting Line1** is
selected automatically.

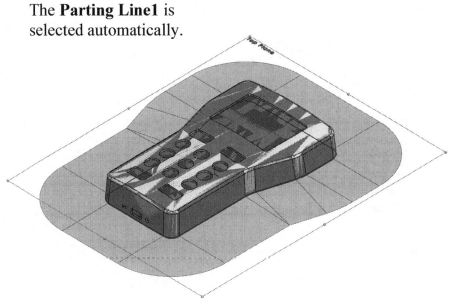

For Parting Surface Distance, enter **3.00in** (arrow).

For Smoothing, use the default **Sharp** option.

Enable the Checkboxes:
Knit All Surfaces and
Show Preview.

Click **OK**.

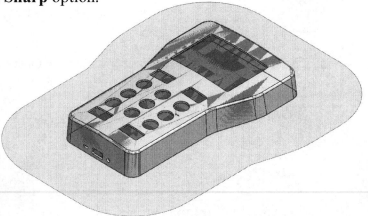

A parting surface is created by
extruding outwards from the parting line.

6. Making the mold block sketch:

The Tooling Split tool uses a sketch to define the final shape and size of the core and cavity mold inserts.

Select the <u>face</u> as noted and open a **new sketch**.

Sketch a **Corner Rectangle** and add dimensions to fully define the sketch.

<u>Exit</u> the Sketch.

Click **Tooling Split** and select the sketch of the **Rectangle**.

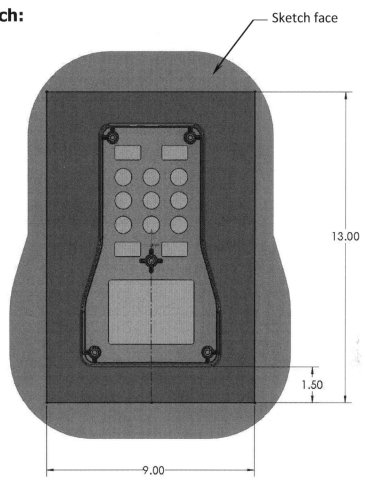

Sketch face

13.00

1.50

9.00

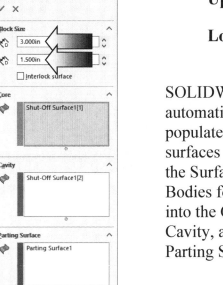

For Block Size, enter the following:

Upper Block = 3.00in.

Lower Block = 1.500in.

SOLIDWORKS automatically populates the surfaces from the Surface Bodies folder into the Core, Cavity, and Parting Surface.

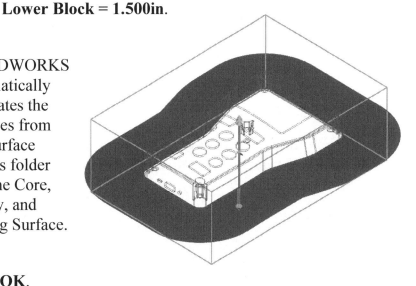

Click **OK**.

7. Analyzing the undercuts:

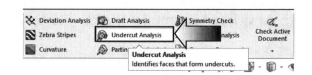

There are trapped areas in this part that cannot be ejected from the mold. The Undercut Analysis is used to find those areas.

Change to the **Evaluate** tab and click **Undercut Analysis** (arrow).

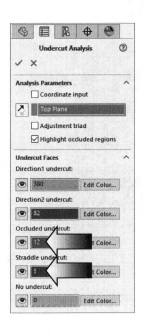

For Direction of Pull, select the **Top** plane.

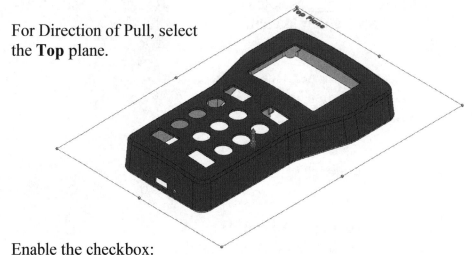

Enable the checkbox:
Highlight Occluded Regions (areas that are not visible from above or below the part).

The following results are displayed:

Direction 1: 380 faces that are not visible from <u>above</u> the part or parting line

Direction 2: 82 faces that are not visible from <u>below</u> the part or parting line

Occluded Undercut: 12 Occluded Undercut Faces (faces not visible from above or below the part)

5 Straddle Faces: 5 Straddle Faces (Faces that draft in both directions)

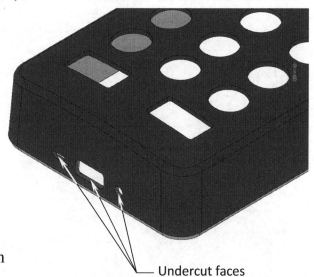

Undercut faces

Click **OK**. A slide core needs to be created to resolve these undercuts.

8. Sketching the slide core profile:

The image on the right shows the 3 holes are <u>not</u> perpendicular to the pull direction. The part cannot be ejected from the mold.

Select the <u>end surface</u> of either the cavity block or the core block and open a **new sketch**.

Sketch face

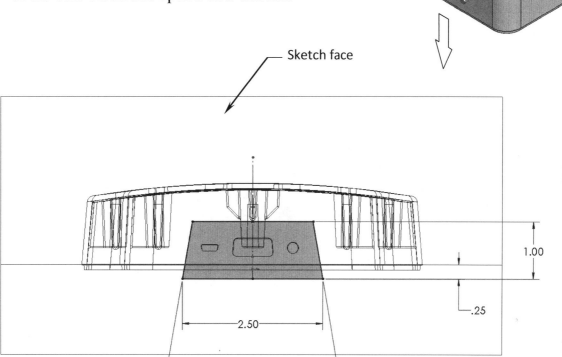

Sketch the profile shown below. Use the mirror function where applicable to maintain the symmetry relations between the lines.

Add dimensions to fully define the sketch.

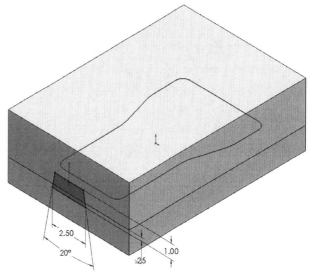

<u>Note</u>: the line at the bottom of the profile is .250" below the cavity block.

9. Extruding the slide core block:

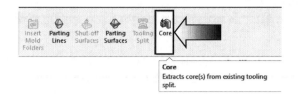

Change to the **Mold Tools** tab.

Click **Core** (arrow).

Under Selections, the **Sketch** of the Core is selected automatically.

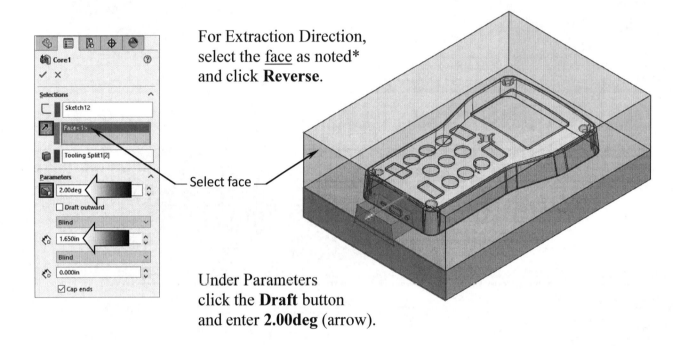

For Extraction Direction, select the <u>face</u> as noted* and click **Reverse**.

— Select face —

Under Parameters click the **Draft** button and enter **2.00deg** (arrow).

For End Condition, use the **Blind** type.

For Extrude Depth, enter **1.650in** (arrow).

Enable the **Cap Ends** checkbox.

Click **OK**.

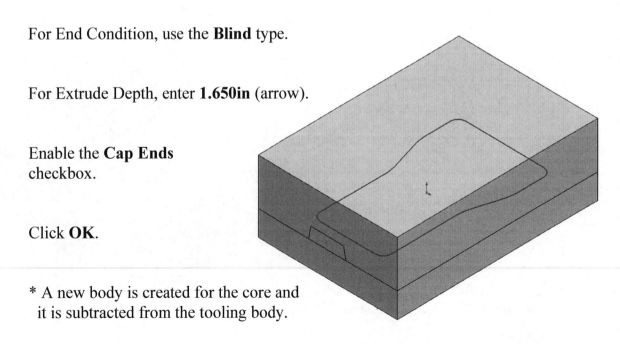

* A new body is created for the core and it is subtracted from the tooling body.

10. Separating the mold blocks:

The Move/Copy Bodies tool is usually used to separate the mold blocks to display their details.

Change to the Features tab and click **Move/Copy** (or select: **Insert, Features, Move/Cop Bodies**).

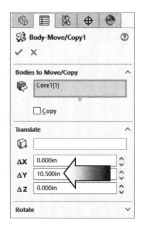

For Bodies to Move/Copy, select the **upper block**.

Click in the Delta Y field and enter **10.500in** (arrow).

Click **OK**.

The upper block moves upward 10.5 inches.
(Notice a small cutout on the bottom of the upper block.)

Click **Move/Copy Bodies** command again.

For Bodies to Move/Copy, select the **lower block**.

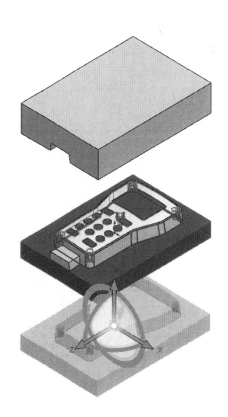

Click in the Delta Y field and enter: **-9.00in.**

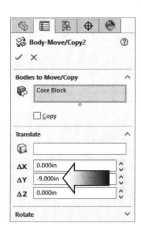

Click **OK**.

The lower block moves downward 9 inches.

The slide core will be moved outward in the next step.

11. Moving the slide core block:

Select the **Move/Copy Bodies** tool one more time.

For Bodies to Move/Copy, select the **Slide Core**.

Click in the <u>Delta Z</u> field and enter **3.00in** (arrow).

Click **OK**.

The Slide Core moves outward
3 inches.

The details of the undercut feature are
now visible for checking or verifications.

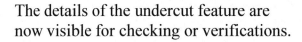

Change to different orientations to verify the details of the slide core.

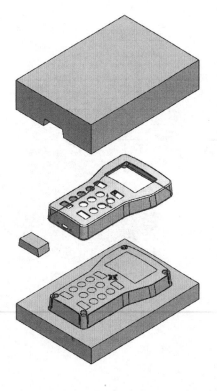

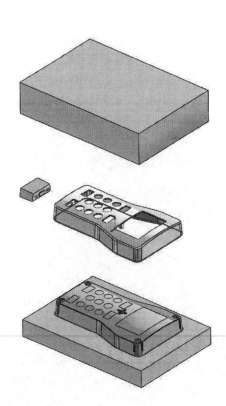

12. Hiding the references:

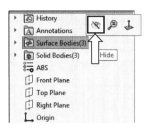

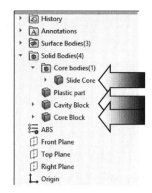

All reference surfaces and parting lines should be hidden at this point.

Click the **Surface Bodies** folder and select **Hide** (arrow).

Expand the Solid Bodies folder and <u>rename</u> the 4 blocks as:

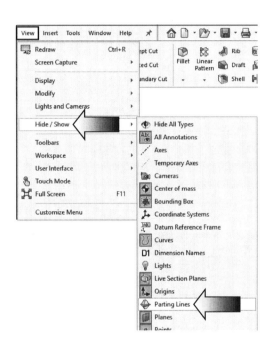

* **Slide Core**

* **Plastic part**

* **Cavity Block**

* **Core Block**

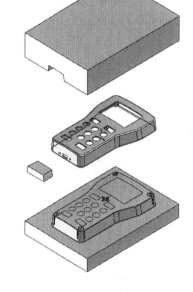

To hide the parting line, select **View, Hide/Show, Parting Lines** (arrow).

Right-click the **Slide Core**, select **Material, Plain Carbon Steel** (arrow).

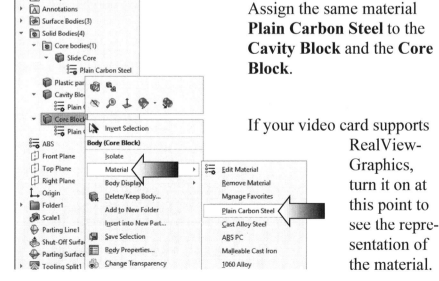

Assign the same material **Plain Carbon Steel** to the **Cavity Block** and the **Core Block**.

If your video card supports RealView-Graphics, turn it on at this point to see the representation of the material.

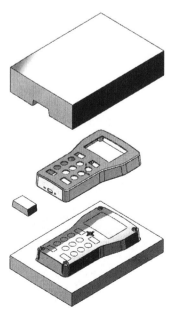

13. Changing the transparency:

To change the transparency of the mold block right-click the **Cavity Block** and select: **Change Transparency** (arrow).

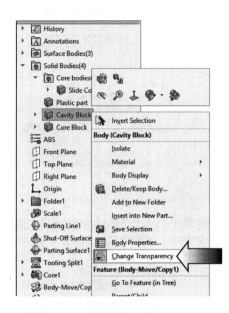

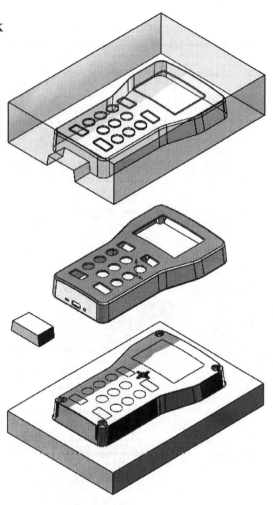

When the exploded view is in use, only the Cavity Block needs to be transparent; but when collapsed, all 3 blocks should be changed to transparent to enhance the visibility of the entire mold.

14. Saving your work:

Select **File, Save As**.

Enter **Undercuts and Slide Core_ Completed.sldprt** for the file name.

Click **Save**.

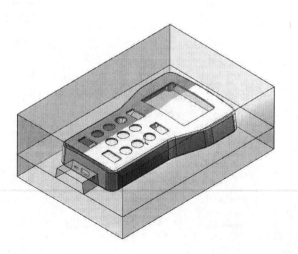

Close all documents.

CHAPTER 9

Alternative Methods

Using Combine Subtract

Sometimes a plastic part does not require a core but instead, 2 cavities are needed to capture the details inside the part. Besides the standard procedures discussed in other lessons, there are other methods to accomplish the desired results such as Combine-Subtract, Cut with Surface, and Assembly Cavity.

The first part of the lesson will teach us those alternative methods to design the mold cavities for the part shown below.

1. Opening a part document:

Select **File, Open**.

Open a part document named:
Alternative Method_Subtract.sldprt

2. Applying the scale factor:

The Scale feature scales only the geometry of the model to accommodate the shrink rate in the mold cavity, it does not scale dimensions, sketches, or reference geometry.

Change to the **Mold Tools** tab and click **Scale**.

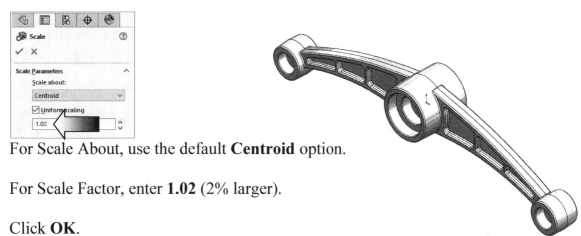

For Scale About, use the default **Centroid** option.

For Scale Factor, enter **1.02** (2% larger).

Click **OK**.

3. Making the mold block sketch:

Select the <u>Front</u> plane and open a **new sketch**.

Sketch a **Corner Rectangle** and add the dimensions shown below.

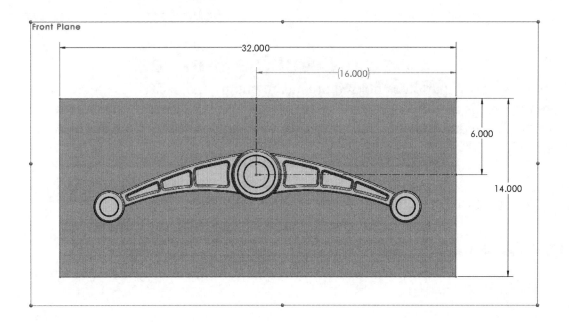

4. Extruding the mold block:

Switch to the **Features** tab and click **Extruded Boss-Base**.

For Direction 1, use the default **Blind** type.

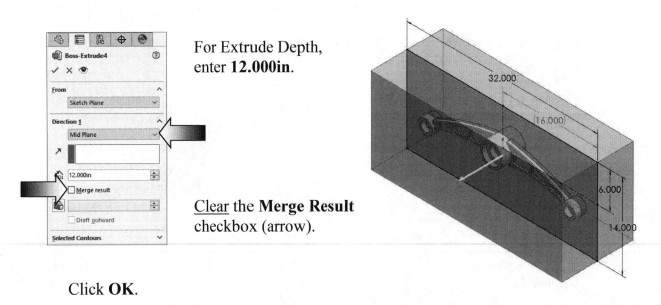

For Extrude Depth, enter **12.000in**.

<u>Clear</u> the **Merge Result** checkbox (arrow).

Click **OK**.

5. Creating a Combine Subtract feature:

The Combine Subtract option uses the geometry of a body to subtract and remove the overlapped material of another body. In this example, the engineered part will be used as the subtract tool to remove the material inside the cavity block.

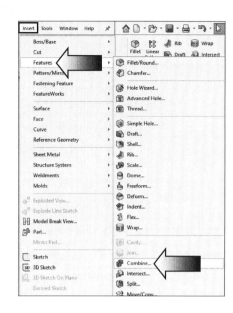

Select **Insert, Feature, Combine** (arrow).

For Operation Type, click **Subtract** (arrow).

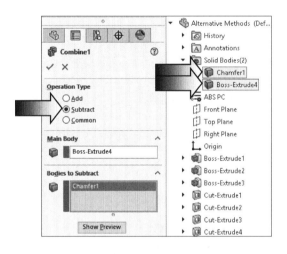

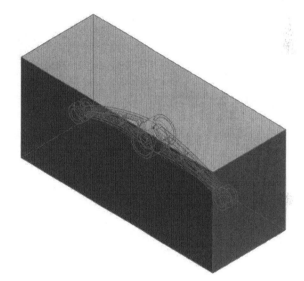

For Main Body, expand the Solid Bodies folder and select: **Boss-Extrude4**, (the extruded block).

For Bodies to Subtract, select:
Chamfer1 (the engineered part).

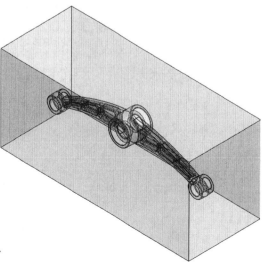

Click **OK**.

Change to Wireframe to see the inside details.

6. Splitting a solid body:

The Split tool is used to split a single solid body into two or more bodies. The split reference can be a line, a surface, or a plane.

From the Features tab, select:
Insert, Features, Split (arrow).

For Trim Tool, select the **Front** plane from the FeatureManager tree.

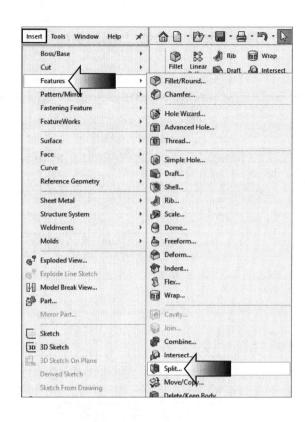

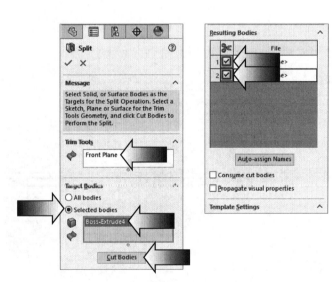

For Target Bodies, click the **Selected Bodies** option and select the **Boss-Extruded1** feature, directly from the graphics area.

Click the **Cut Bodies** button (arrow).

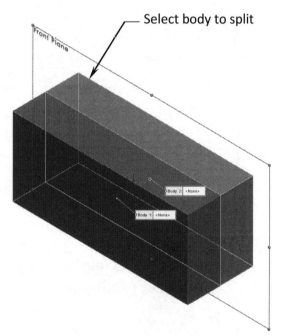

Select body to split

Enable the **2 checkboxes** under the Resulting Bodies section (arrow).

Click **OK**.

7. Rotating a solid body:

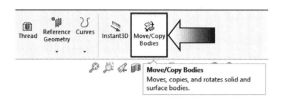

Click **Move/Copy Body** (or select:
Insert, Features, Move/Copy).

For Bodies to Move/Copies, select the **solid body** in the <u>front</u>.

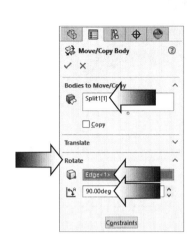

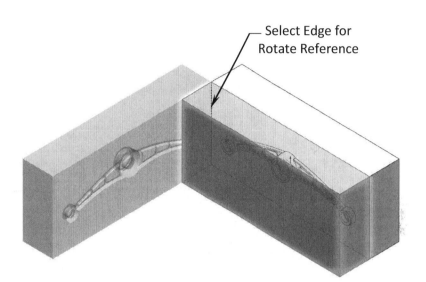

Select Edge for
Rotate Reference

Expand the Rotate section and select the **Edge** as indicated, for Rotate Reference.

For Rotate Angle, enter **90.00deg**.

The front cavity block should rotate outward, similar to the image shown below.

Click **OK**.

<u>Note:</u> The engineered part is consumed by the subtract operation. Use the Move/Copy option to make a copy of the engineered part before creating the combine-subtract feature, if needed.

8. Saving your work:

Select **Files, Save As**.

Enter: **Alternative
Methods_Subtract**
for the file name.

Click **Save**.

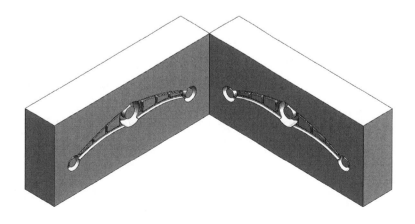

Using Cut with Surfaces

This 2nd part of the lesson will teach us another alternative method to create the cavity mold for the same part by using surfaces.

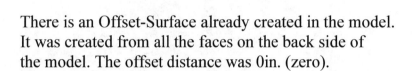

1. Opening a part document:

Close the previous part document and <u>open</u> another part document named: **Alternative Methods_Surface.sldprt**

There is an Offset-Surface already created in the model. It was created from all the faces on the back side of the model. The offset distance was 0in. (zero).

2. Isolating the surface body:

Right-click the **Surface Bodies** folder and select **Isolate** (arrow).

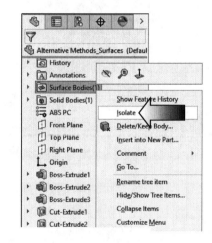

The solid model is hidden; only the surface body is visible in the graphics area.

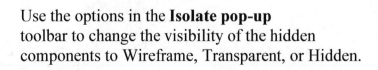

Use the options in the **Isolate pop-up** toolbar to change the visibility of the hidden components to Wireframe, Transparent, or Hidden.

Keep the option at **Hidden** for the next few steps.

3. Making the mold block sketch:

Select the <u>Front</u> plane and open a **new sketch**.

Sketch a **Corner Rectangle** around the surface body and add the dimensions shown in the image to fully define the sketch.

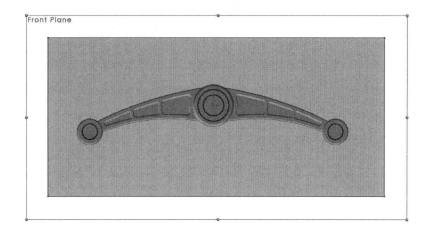

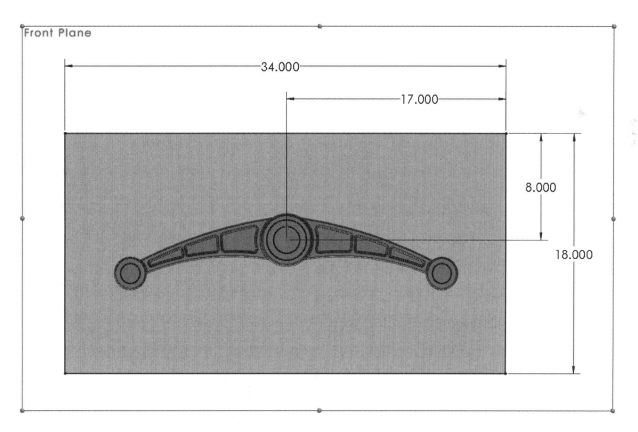

Do not exit the sketch; keep it active for the next step.

4. Converting to planar surface:

Change to the **Surfaces** tab.

Click **Planar Surface** (arrow).

For Bounding Entities, select the **current sketch** (the rectangle).

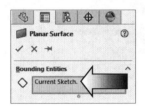

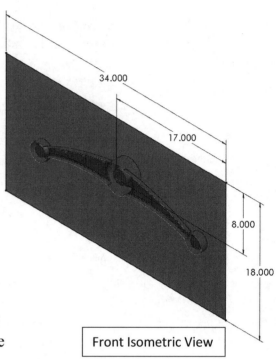

Front Isometric View

The preview graphics shows a planar surface is being created from the sketch.

Click **OK**.

To precisely view the front and the back sides of the isometric view, use the following shortcuts:

* **Control + 7 = Front Isometric View**

* **Shift + Up Arrow key twice = Back Isometric view**

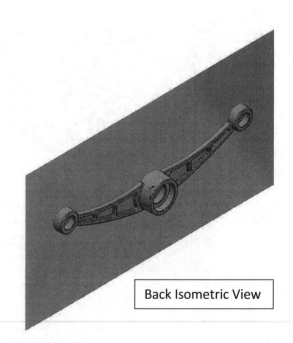

Back Isometric View

(Take a moment to practice the shortcuts because we will be toggling back and forth between the front and back isometric views a few times in this lesson.)

5. Trimming the surfaces:

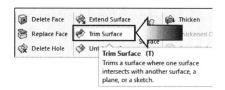

Switch to the **Surfaces** tab.

Click **Trim Surface**.

For Trim Type, select **Mutual** (arrow).

For Selections, select the **Planar Surface** and the **Offset-Surface** from the graphics area.

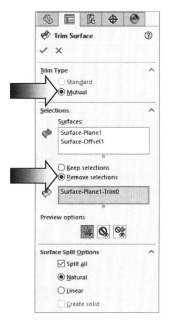

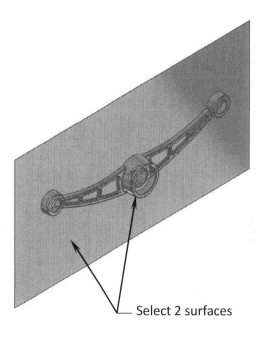

Select 2 surfaces

Click the **Remove Selections** option and select the center portion of the planar surface as noted.

Click **OK**.

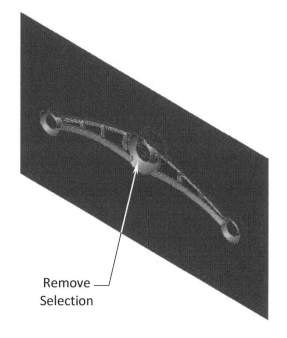

Remove Selection

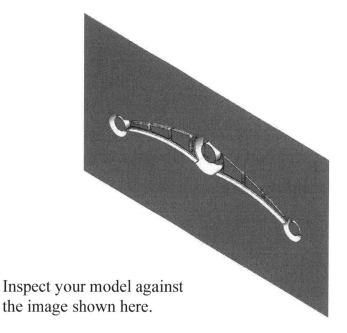

Inspect your model against the image shown here.

6. Knitting the surfaces:

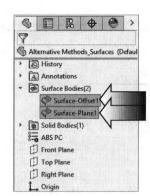

There are 2 surfaces in the Surface Bodies folder.
They should be knitted into a single surface so that
it can be used as the trimming surface in the next steps.

Click **Knit Surface**.

For Selections, select the **2 surfaces** either directly from
the graphics area or from the Surface Bodies folder.

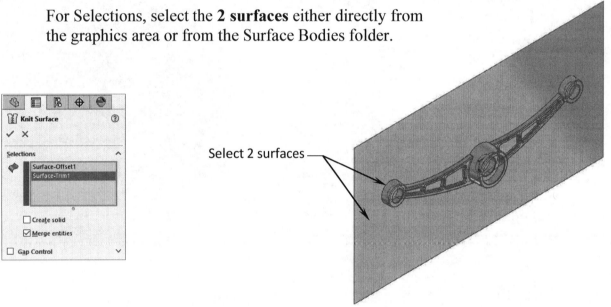

Select 2 surfaces

Enable the **Merge Entities** checkbox.

Click **OK**.

The 2 surfaces are
knitted into a single
surface.

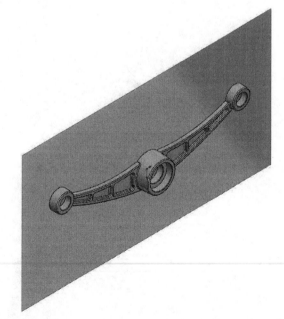

The Surface Bodies
folder shows only 1
surface inside (arrow).

7. Sketching the profile of the mold block:

Click **Exit Isolate** to show the solid part again.

Select the <u>Front</u> plane and open a **new sketch**.

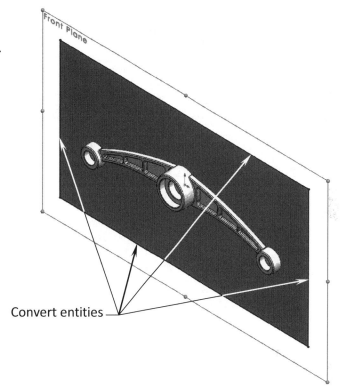

Select the <u>4 edges</u> of the planar surface and click **Convert Entities**. This creates an On-Edge relation between each edge and line. If the planar surface changes to a different size, the converted lines will also change.

Convert entities

Change to the **Features** tab.

Click **Extruded Boss-Base**.

For Direction 1, use the default **Blind** type.

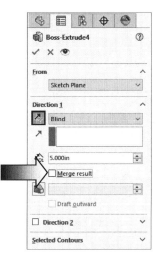

For Extrude Depth enter: **5.000in**.

Click the **Reverse Direction** arrow.

<u>Clear</u> the **Merge-Result** checkbox.

Click **OK**.

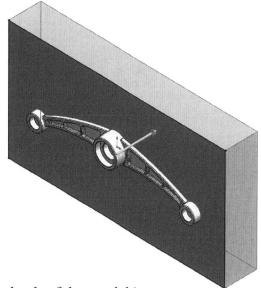

(The extruded thickness should be added to the back of the model.)

8. Cutting with a surface:

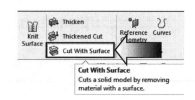

A surface can be used to cut a solid body. This is similar to an extruded cut feature except for the cut feature is made with a surface.

Change to the Surfaces tab and click: **Cut with Surface** (arrow).

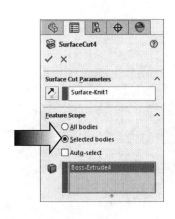

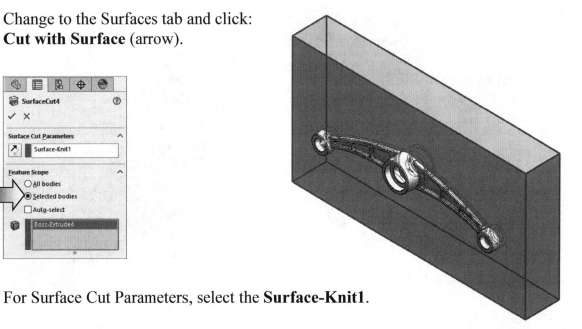

For Surface Cut Parameters, select the **Surface-Knit1**.

For Feature Scope, click Selected Bodies and select the **Boss-Extrude4** feature.

Click **OK**.

Select body to cut

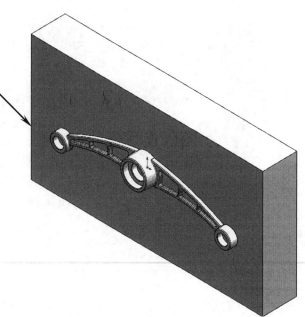

The Knit Surface removes the material that intersects it, leaving a cavity behind it.

We will move the solid part to examine the cavity cut inside the mold block.

9. Moving a solid body:

Click **Move/Copy Bodies** (or select: **Insert, Features, Move/Copy**).

For Bodies to Move/Copy, select the **engineered part**.

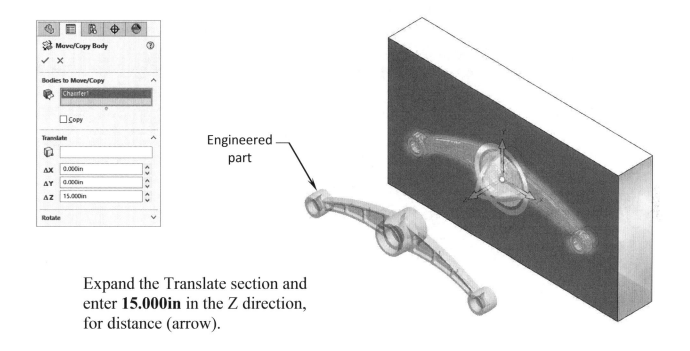

Engineered part

Expand the Translate section and enter **15.000in** in the Z direction, for distance (arrow).

Click **OK**.

10. Saving your work:

Select **File, Save As**.

Enter: **Alternative-Methods_Surfaces.sldprt** for the file name.

Click **Save**.

Close the part document.

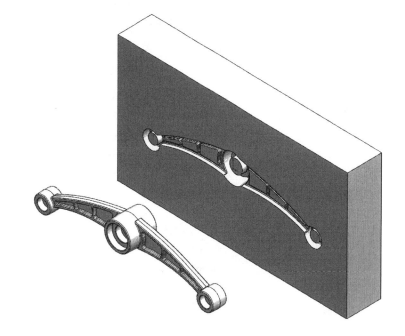

Using the Cavity Feature in Assembly

An alternate approach to create simple molds is to use the cavity tool. A cavity in the shape of the designed part is created in the mold base part. The cavity size reflects the scaling factor specified in the part.

1. Opening an assembly document:

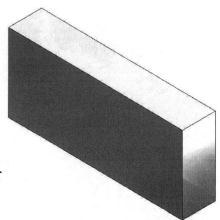

Open an assembly document named:
Alternative Methods_Assembly.sldasm

This assembly document contains 1 component. The engineered part will be inserted into the center and used to subtract a cavity from it. The scale factor is entered during the cavity creation.

2. Inserting the engineered part:

Change to the **Assembly** tab.

Select **Insert Component** (arrow).

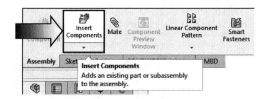

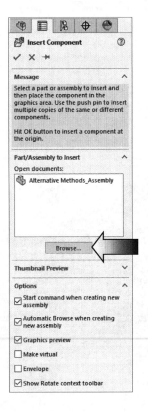

Click the **Browse** button and select the component named:
Alternative Methods_Assembly.sldprt

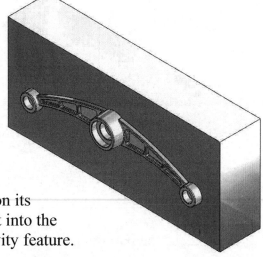

Click **OK** (the **green check mark**) to place the component on the assembly's origin.

The new component is placed in the center of the mold block. The geometry on its back side will be used to cut into the mold block to create the cavity feature.

3. Editing a component:

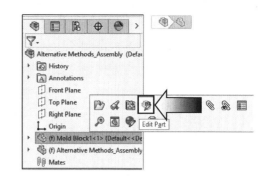

The mold block must be active in order to add a cavity cut feature to it. Select the component **Mold Block1** and click **Edit Component** (arrow).

The selected component changes into the blue color. It can now be edited.

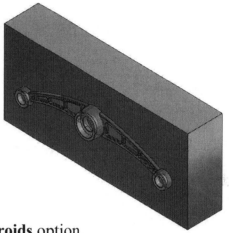

Select: **Insert, Features, Cavity** (arrow).

For Design Components, select the component **Alternative Methods_Assembly** from the FeatureManager tree.

For Scale About, use the default **Component Centroids** option.

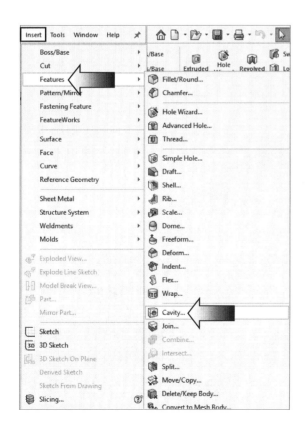

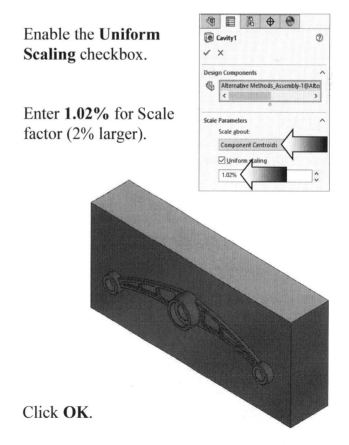

Enable the **Uniform Scaling** checkbox.

Enter **1.02%** for Scale factor (2% larger).

Click **OK**.

A cavity, which is 2% larger than the engineered part, is created in the Mold Block1 component.

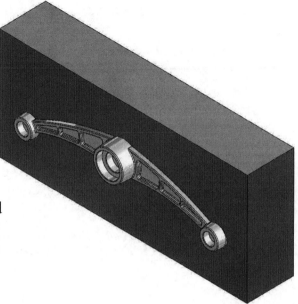

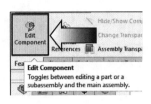

Click-off the **Edit Component** command to return to the Assembly Edit mode. An exploded view is needed to be able to see the cavity feature.

4. Creating an exploded view:

Click **Exploded View** (arrow).

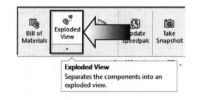

Select the **engineered part** (the Alternatives Methods_ Assembly component).

Drag the arrowhead of the Z direction outward approximately **15.00 inches**.

Click **OK**.

Zoom in closer to examine the cavity feature.

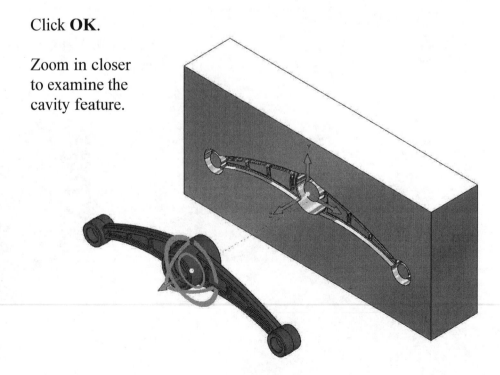

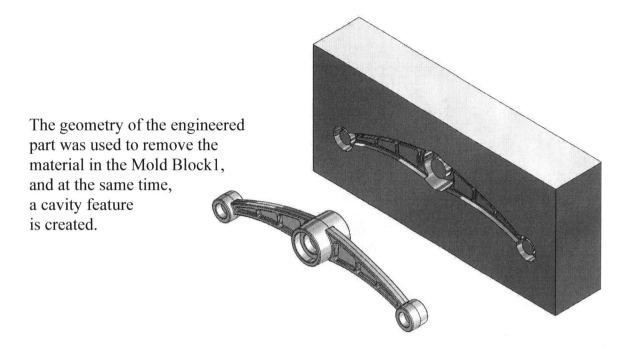

The geometry of the engineered part was used to remove the material in the Mold Block1, and at the same time, a cavity feature is created.

5. Saving your work:

Select **File, Save As**.

Enter: **Alternative Methods_Assembly_Completed.sldasm** for the file name.

Click **Save**.

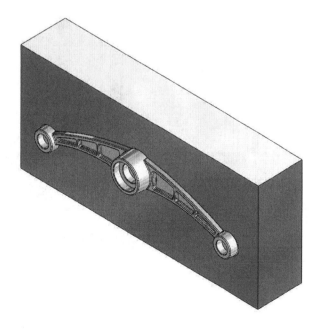

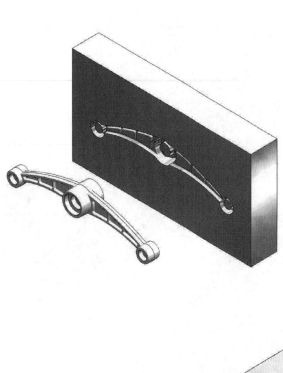

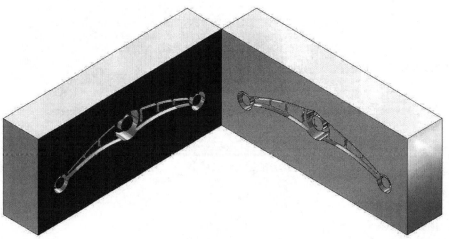

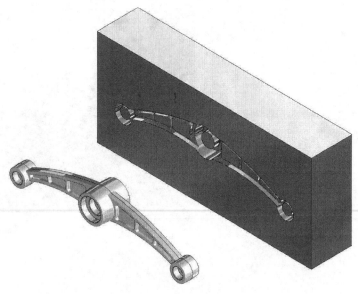

CHAPTER 10

Thickness Analysis

The Thickness Analysis utility is used to find thick and thin regions of a plastic part. This tool can also assist in the design of injection moldings and identify potential failure regions or design flaws.

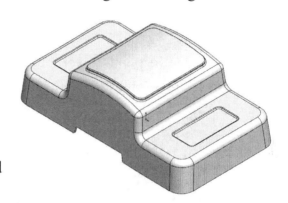

1. Opening a part document:

Select **File, Open**.

Open a part document named:
Thickness Analysis.sldprt

The material **ABS** has already been assigned to this part.

2. Setting the parameters:

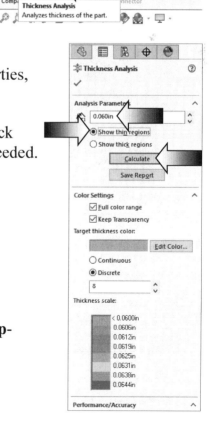

The Thickness Analysis utility determines the different thicknesses of a thin-walled plastic part. It can also generate a report that includes Summary, Analysis Details, Mass Properties, and Thickness Scale in any of the Model Views.

This lesson will show us how to find the Thin and Thick regions in the part so that corrections can be done if needed.

Change to the **Evaluate** tab and click:
Thickness Analysis (arrow).

Under Analysis Parameters, for **Target Thickness**, enter **.060in.** (arrow).

Select the **Show Thin Regions** option (arrow).
Under **Color Settings,** select: **Full Color Range**, **Keep-Transparency**, **Discrete-8** color bands.

Click: **Calculate**.

3. Analyzing the thickness:

The Thickness Analysis utility runs the analysis based on the parameters provided. The part is displayed in various shades to indicate the different thicknesses.

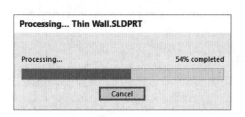

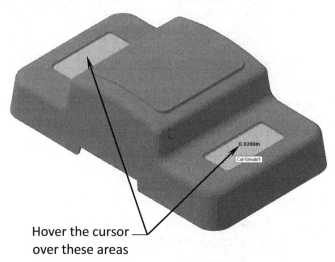

Hover the pointer over the part to see the tooltip display thickness values.

Hover the cursor over these areas

The **2 recessed features** are much thinner than the rest. The thickness in these areas is only about **.020 inches**.

Click the Show **Thick Regions** option (arrow) and set the **Thick Region Thickness** to **.065in**.

Click **Calculate**.

Hover the cursor over the center of the part. The thickness in this area is about **.140 inches**. It is much thicker than other areas.

Hover the cursor over this area

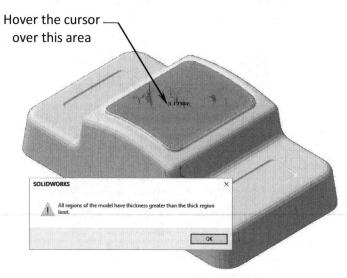

4. Creating a section view:

Section View
Displays a cutaway of a part or assembly using one or more cross section planes.

An alternative option is to create a section view and inspect or measure the different thicknesses in the part.

Click **Section View** on the View Heads-Up toolbar (arrow).

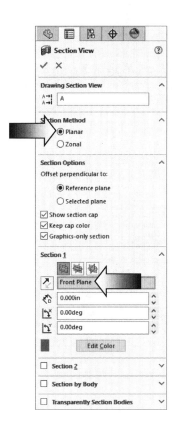

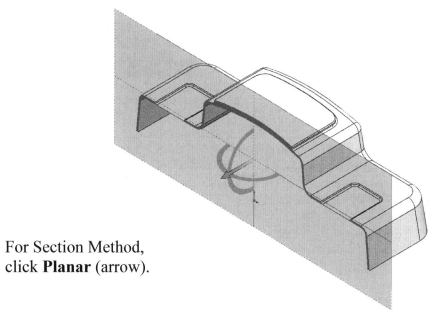

For Section Method, click **Planar** (arrow).

For Cutting Plane, select the **Front** plane (arrow).

Click **OK**.

The default thickness of the part is .060in. The section view shows a couple of very thin regions (.020in.), and a very thick region (.140in.). Ideally, we want a constant wall of .060in throughout the part.

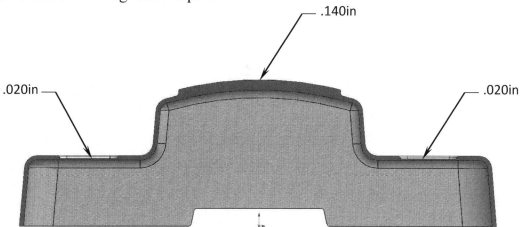

.140in

.020in

.020in

5. Re-arranging the shell feature:

One of the quick solutions to this issue is to rearrange the features from the Feature Manager tree.

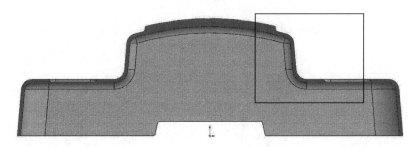

By moving the Shell feature below the Cut, Boss, Fillet, and Draft features, we can change the thickness of the part to one constant wall.

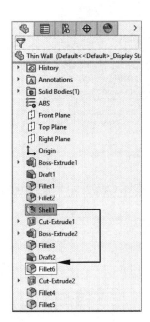

Drag the **Shell1** feature down the tree and drop it right <u>below</u> the **Draft2** feature (arrow).

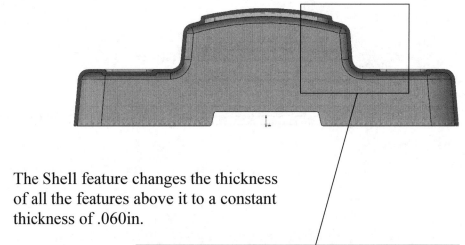

The Shell feature changes the thickness of all the features above it to a constant thickness of .060in.

The Shell feature is usually added towards the end of the plastic part development process, so that all features that were created before the shell feature can have the same thickness throughout.

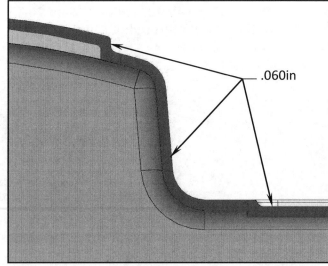

.060in

<u>Exit</u> the Section View command.

6. Scaling the part:

Change to the **Mold Tools** tab and click: **Scale**.

For Scale About, select **Centroid**.

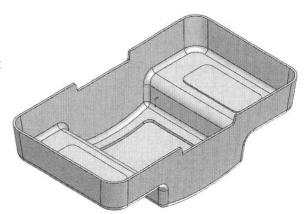

Enable the **Uniform Scaling** checkbox.

For Scale Factor, enter **1.02** (2% larger).

Click **OK**.

7. Creating the parting lines:

Click **Parting Lines** (arrow).

For Direction of Pull, select the **Top** plane from the Feature-Manager tree.

For Draft Angle, enter **2.00deg**.

Click **Draft Analysis** (arrow).

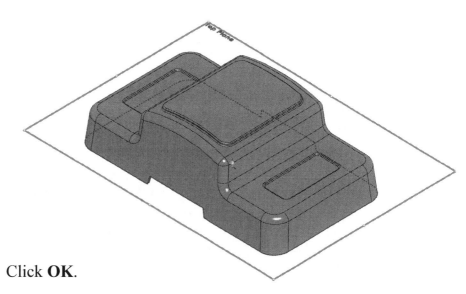

Click **OK**.

8. Creating a parting surface:

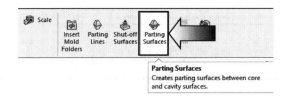

There is no through hole in this part,
so the Shut-Off surfaces can be omitted.

Click **Parting Surfaces** (arrow).

For Mold Parameters, select **Perpendicular to Pull** (arrow).

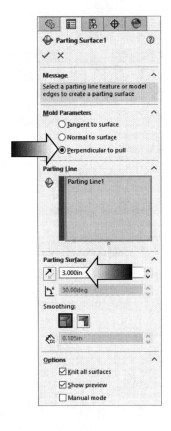

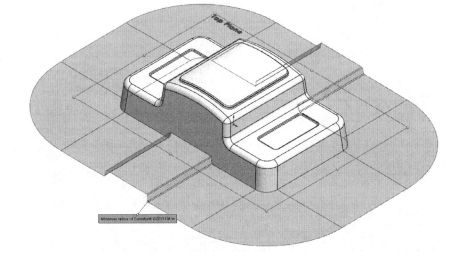

The **Parting Line1** is selected automatically.

For Parting Surface Distance, enter **3.00in**. (arrow).

For Smoothing,
use the default
Sharp option.

Enable the check-
boxes as shown.

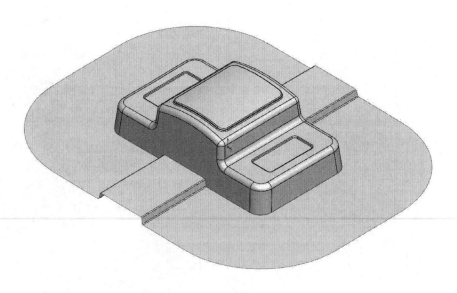

Click **OK**.

9. Adding a new plane:

Change to the **Features** tab and select: **Reference Geometry, Plane** (or select: **Insert, Reference Geometry, Plane**).

For First Reference, select the **Top** plane.

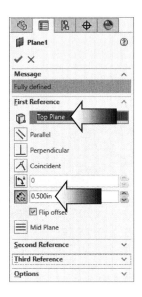

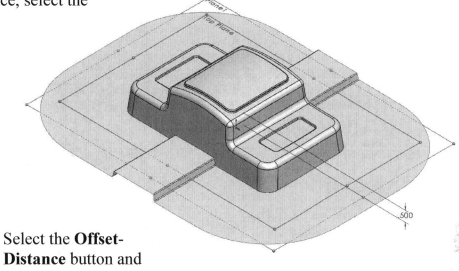

Select the **Offset-Distance** button and **.500in.** (arrow).

Click the **Flip Offset** checkbox to place the new plane <u>below</u> the Top plane.

Click **OK**.

10. Making the mold block sketch:

Open a **new sketch** on the new <u>Plane1</u>.

Sketch a **Corner-Rectangle** and **two Centerlines**.

Add the **Midpoint** relations as indicated.

Add dimensions to fully define the sketch.

<u>Exit</u> the sketch.

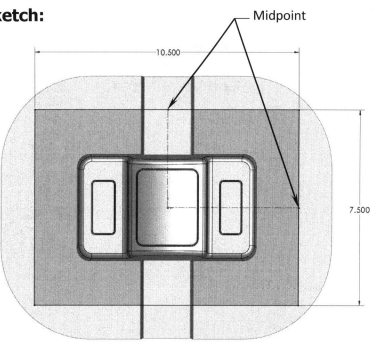

11. Extruding the mold blocks:

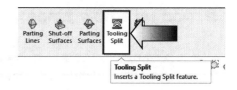

Change to the **Mold Tools** tab.

Select the sketch of the **Rectangle** and click **Tooling Split** (arrow).

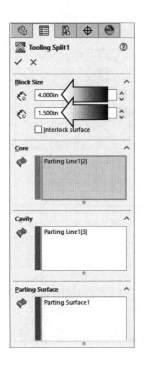

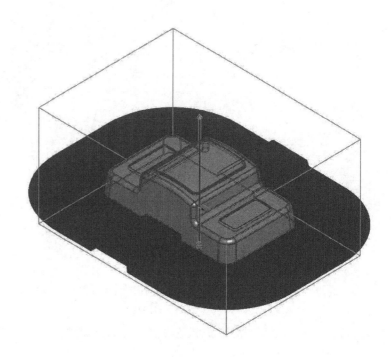

For Block Size, enter the following:

Upper Block = 4.00in.

Lower Block = 1.500in.

The surfaces in the Surface Bodies folder are populated automatically into the Core, Cavity, and Parting Surface sections.

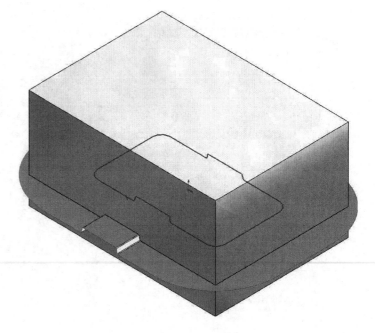

Click **OK**.

The 2 mold blocks are created.

12. Hiding the references:

The reference surfaces and parting lines are usually put away at this point.

To hide the parting lines, select **View, Hide/Show, Parting Lines** (arrow).

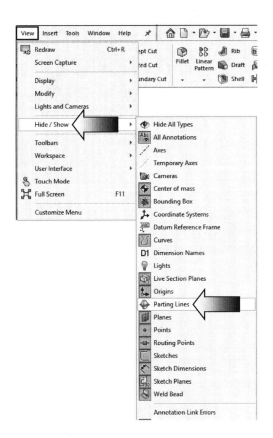

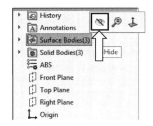

Click the **Surface Bodies** folder and select: **Hide** (arrow).

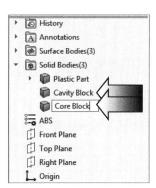

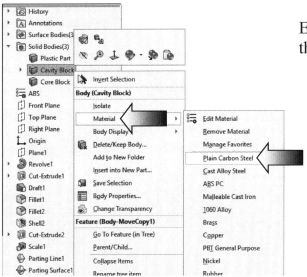

Expand the Solid Bodies folder and <u>rename</u> the 3 solid bodies to:

* **Plastic Part**
* **Cavity Block**
* **Core Block**

Right-click the **Cavity Block** and select **Material, Plain Carbon Steel**.

Apply the same material to the Core.

13. Separating the mold blocks:

Change to the **Features** tab and click **Move/Copy** (or select: **Insert, Features, Move/Copy**).

For Bodies to Move/Copy, select the **upper block** (the Cavity).

Click in the <u>Delta Y</u> field and enter **8.500in**.

Click **OK**.

The upper block moves upward 8.5 inches.

Click **Move/Copy Bodies** once again.

For Bodies to Move/Copy, select the **lower block** (the Core).

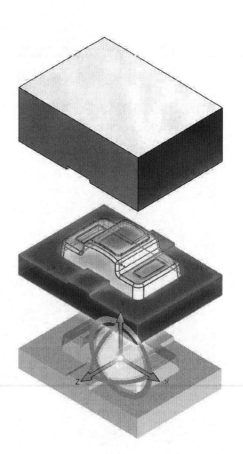

Click in the <u>Delta Y</u> field and enter **-7.000in**. (arrow).

Click **OK**.

The lower block moves downward 7 inches.

Change to different orientations to inspect the details of the Core and Cavity blocks.

14. Changing the transparency:

Expand the **Solid Bodies** folder.

Right-click the Cavity Block and select:
Change Transparency (arrow).

The default transparent
amount is 75%.

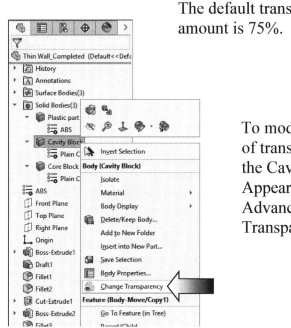

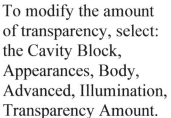

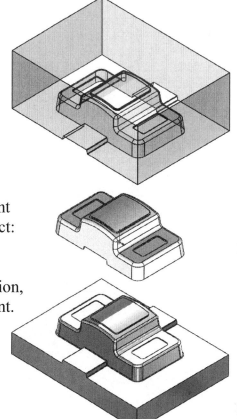

To modify the amount
of transparency, select:
the Cavity Block,
Appearances, Body,
Advanced, Illumination,
Transparency Amount.

NOTE: When the explode view is active, only the Cavity Block
needs to be shown as transparent; but when it is collapsed, the Core Block
may also need to be changed to transparent so that the interior details will still
be visible without the need of creating the additional views.

15. Saving your work:

Select **File, Save As**.

Enter **Thickness Analysis_Completed.sldprt** for the file name.

Click **Save**.

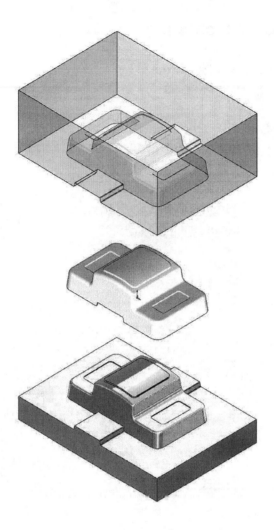

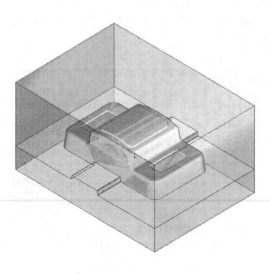

CHAPTER 11

Using SOLIDWORKS Plastics

SOLIDWORKS Plastics Standard,
SOLIDWORKS Plastics Professional, and
SOLIDWORKS Plastics Premium are available as separately purchased products that can
be used with SOLIDWORKS Standard, SOLIDWORKS Professional, and SOLIDWORKS
Premium.

SOLIDWORKS Plastics simulates how melted plastic flows during the injection
molding process to predict manufacturing related defects on parts and
molds. A new study is created to define the key characteristics of
your simulation, specifically whether you are simulating single
or multiple materials, and whether you are choosing the solid
or shell analysis procedure.

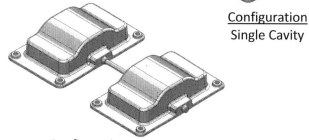

Configuration
Single Cavity

1. Opening a part document:

Select **File, Open**.

Open the part document named:
Basin.sldprt

Configuration
Double Cavity

2. Enabling SOLIDWORKS Plastics:

This part has 2 configurations, Single Cavity
and Double Cavity.

The Double Cavity will take a lot more time
to analyze. The Single Cavity will be used
and it is the active configuration.

To enable SOLIDWORKS Plastics, select:
Tools, Customize and click the
SOLIDWORKS Plastics checkbox.

Click **OK**.

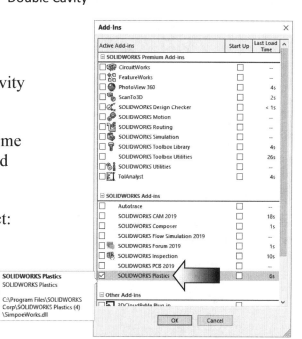

3. Creating a new study:

Switch to the **SOLIDWORKS Plastics** tool tab (arrow).

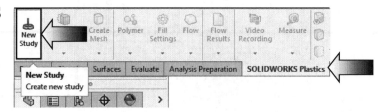

Click **New Study** (arrow).

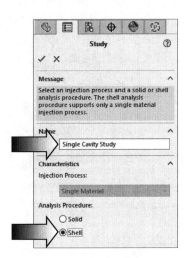

For Study Name, use the default **Single Cavity Study**.

For Analysis Procedure, select **Shell**.

The shell analysis procedure can only support a single domain of the Cavity type. The shell procedure cannot accurately model grid-like or strip-like configurations, or parts with step-changes in thickness.

Click **OK**.

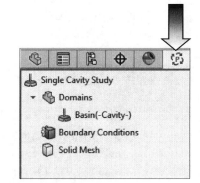

The **PlasticsManager** becomes available on the tree.

The three nodes: **Domains**, **Boundary Conditions** and **Shell Mesh** are displayed on this tree.

A Domain represents a volume in space through which you simulate a heat or fluid flow. For the analysis to proceed, you must assign a Domain type to each body or exclude from the analysis the bodies without a Domain type. With the shell analysis procedure, the only supported Domain type for a solid body is cavity.

4. Specifying the injection location:

Melted plastic material flows into the part cavity through injection locations.

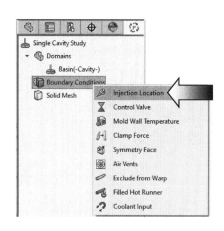

The Injection Location introduces polymer material at the specified melt temperature into the cavity.

Right-click **Boundary Conditions** and select: **Injection Location**.

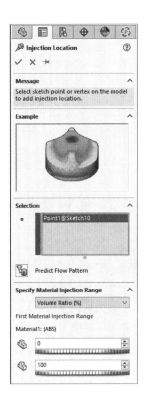

Select point

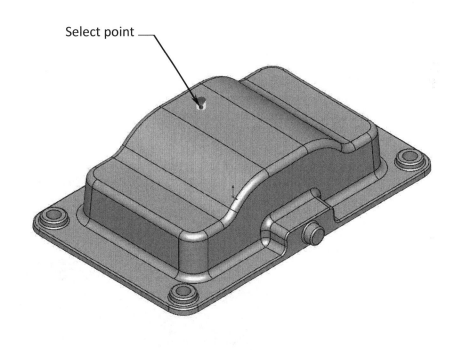

Change to the Isometric view (Control+7).

For Face Selection, select the **Sketch Point** as indicated.

Click **OK**.

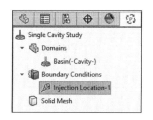

The Injection Location is created and displayed on the tree.

5. Creating a Shell Mesh:

Shell analysis is the most computationally efficient for thin walled parts with moderate thickness variations and no narrow flow channels.

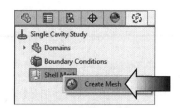

Right-click **Shell Mesh** and select **Create Mesh**.

For Surface Mesh, use the <u>default</u> Mesh Density.

For Refinement Method, select **Curvature-Based**.

Select **Curvature-based** local refinement to refine the mesh near the holes, the injection location, and the corners. Refining the mesh in critical regions improves the accuracy of results but increases the computational time.

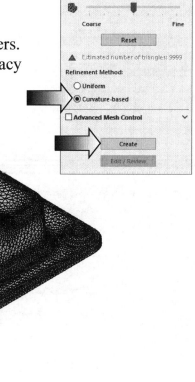

Click **Create**.

A surface mesh with the current settings is created and displayed in the graphics area. The Create button is activated again when you modify the current mesh settings.

Click **OK**.

6. Assigning material:

Expand the **Material** option on the PlasticsManager tree.

Right-click **Polymer** and select **Open Database**.

Select the **Default Database** option (arrow).

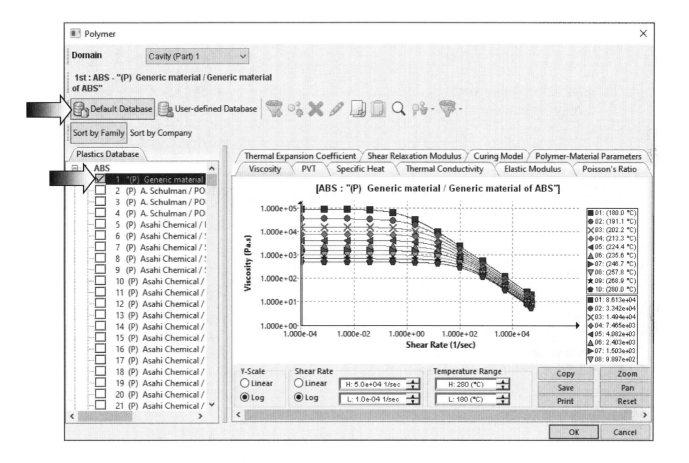

Under Plastics Database, expand **ABS** and click the first checkbox for:
Generic material of ABS.

Click **OK**.

The Polymer icon now has a green check mark, indicating a material has been assigned to the part.

7. Running the flow analysis:

This step predicts the filling pattern of the material melt flow, as it fills the cavity.

In the PlasticsManager tree, expand **Run**.

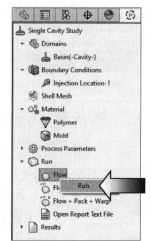

Right-click Flow and select **Run** to start the flow analysis.

A progress bar appears showing the status of the Flow analysis. It may take a few minutes to complete the analysis.

After the Flow analysis is completed, the **Results** Property-Manager lists the plots under the **Available Results** section.

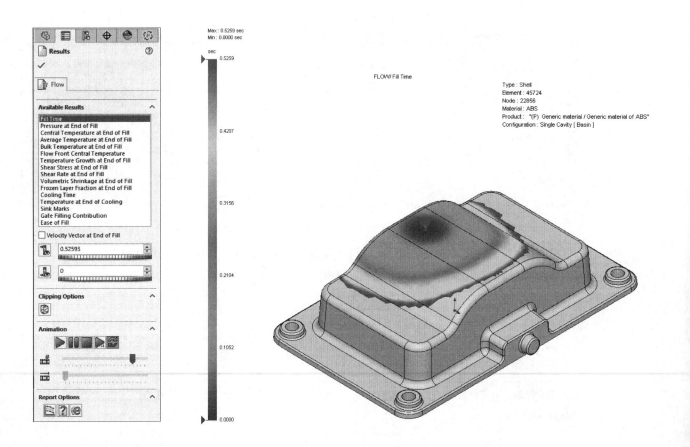

8. Viewing the results:

The first available result is the **Fill Time** plot.

The results show the time when each runner and cavity domain is filled.

Click **Play Animation** to view the cavity is being filled with the material.

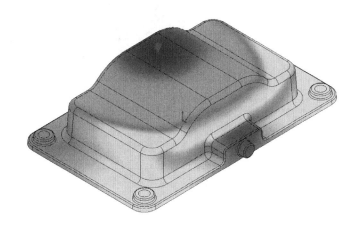

The **Advisor** button is automatically enabled to display the **Results Adviser** on the right-hand side of the screen.

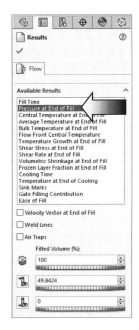

The Pressure at End of Fill Plot

Select other Flow Results such as **Temperature at End of Cooling** or **Ease of Fill**.

Review the associated descriptions in the **Results Adviser** window.

The Temperature at End of Cooling Result Plot

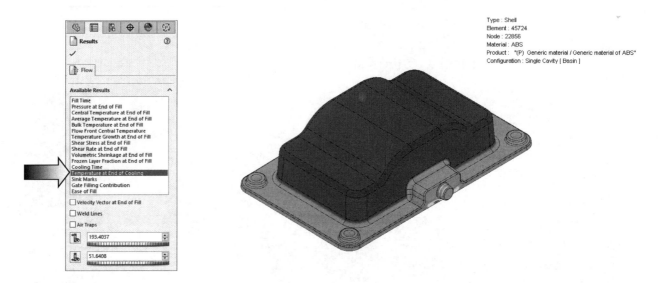

The Advisor indicates this part can be successfully filled with an injection pressure of 49.8 MPa (7230.64 psi). The injection pressure required to fill is less than 66% of the maximum injection pressure limit specified for this analysis, which means you are well under your specified limit.

The Ease of Fill Result Plot

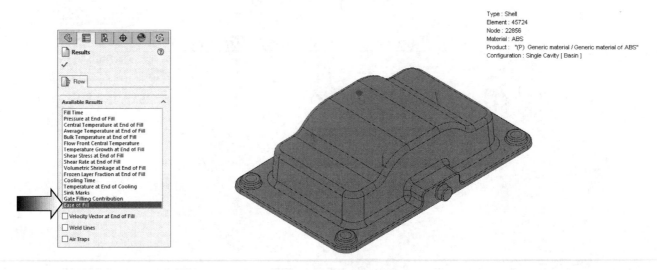

The result has a uniform green color indicating that the cavity fills successfully based on the geometry, the injection location, and the default process parameters.

Click **OK** to exit the Results.

9. Viewing the report text file:

You can view a summary of the analysis and generate a project report to communicate key findings and results of the simulation to the team members.

To open the Summary and Report PropertyManager, do one of the following:

* In the PlasticsManager tree, expand Results, and double-click **Summary and Report**.
or
* In the Plastics CommandManager, click **Generate Report**.

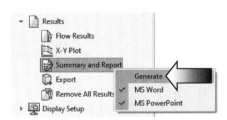

A text file of the report is also generated and stored under **Run / Open Report Text File**.

Expand **Open Report Text File**.

Right-click **Flow Text** and select **Open**.

Internet Explorer is used to display the report.

The report includes the information regarding the analysis such as: Solver Parameter Information, Process Information, Material Data Information, Geometric Data Information, and Filling Stage Results Summary.

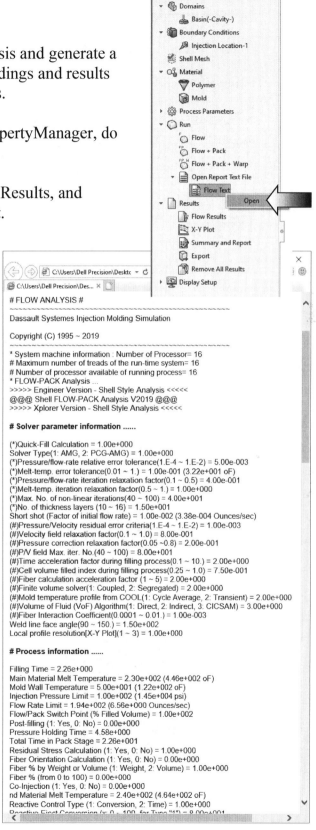

10. Saving your work:

Select **File, Save As**.

Enter: **Basin_Completed.sldprt** for the file name.

Click **Save**.

SOLIDWORKS saves information such as the details of the study, the domain assignment, boundary conditions such as injection location and mesh parameters, with the part document.

The Plastics analysis settings and results are saved in the same location as the part file.

Close all documents.

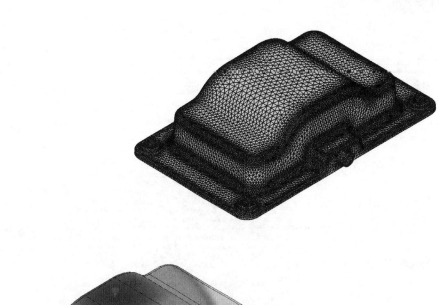

CHAPTER 12

Plastics Flow Analysis

When a plastic part is made by injection molding,
the plastic pellets are loaded into a hopper. They get heated and melted into liquid resin and
then forced into a cavity to fill it (Fill Stage). This is when the shrinkage starts to take place.
To minimize the shrinking, additional liquid resin is forced into the cavity under constant
pressure (Pack Stage). The plastic starts to solidify in the mold, and within a few seconds
when it reaches the ejection temperature (Cool Stage), it is then ejected from the mold.

This lesson will walk us through the 5 basic steps of performing the analysis: Injection
Location, Mesh, Material, Run and View the analysis results.

1. Opening a part document:

Click **File, Open**.

Open a part document named:
Plastics_Flow Analysis.sldprt

(This model has 2 configurations:
With Runner & Gate and Without
Runner & Gate.
The Without Runner & Gate
is the active configuration.)

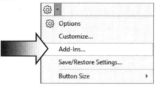

2. Enabling Plastics:

Click the drop-down arrow next to the
gear symbol (Options) and select:
Add-Ins (arrow).

Under SOLDWORKS Add-Ins, enable
the checkbox for **SOLIDWORKS-Plastics** (arrow).

Two new tabs: **SOLIDWORKS Plastics**
and **Analysis Preparation** are added to the CommandManager.

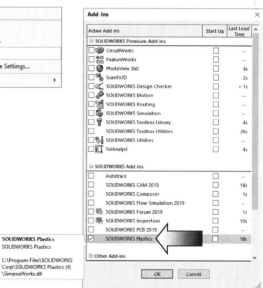

3. Setting up the Mesh:

Change to the SOLIDWORKS **Plastics** tab and select: **New Study** (arrow).

For Study Name, enter: **Flow Analysis**.

For Analysis Procedure, select: **Shell** (arrow).

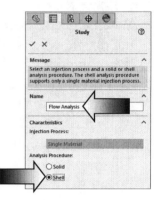

Click **OK**.

A PlasticsManager tree appears on the left side of the screen.

4. Viewing the PlasticsManager tree:

The PlasticsManager tree contains the information regarding the simulation study such as:

> *** Domains**
> *Represents a volume in space through which you simulate a heat or fluid flow.*

> *** Boundary Conditions**
> *Settings for specifying the injection location.*

> *** Shell Mesh**
> *The shell analysis procedure requires a shell mesh, which is appropriate for thin-walled parts with uniform thickness.*

Several other tools such as Material, Process Parameters, Run, and Results will be added to the PlasticsManager tree after the Shell Mesh is created.

5. Adding an Injection Location:

Injection locations introduce polymer material at the specified melt temperature into the cavity.

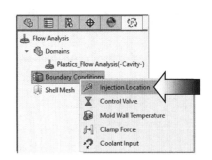

Expand the **Boundary Condition** feature and double click on **Injection Location** (arrow).

For injection location, select the <u>sketch point</u> as noted (3D Sketch1).

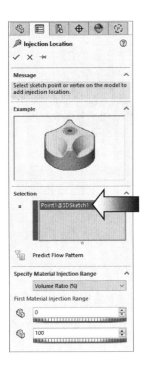

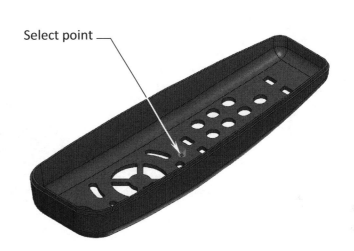

Select point

(A red conical pointer appears at the injection location. Its diameter can only be changed using the solid analysis procedure.)

Click **OK**.

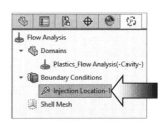

The Injection Location is captured and saved under the Boundary Conditions section.

6. Creating a mesh:

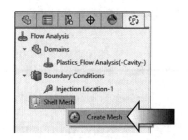

A mesh subdivides each domain of the simulation model, the cavity, runner system, cooling channel, inserts, and mold, into discrete cells.

Within each cell, SOLIDWORKS applies the appropriate conservation equations. The conservation equations compute the flow of melted polymer and heat, simulate phase change as the melt cools, and predict residual stresses and their effect on the unconstrained part shape.

For Surface Mesh, use the **default** mesh density.

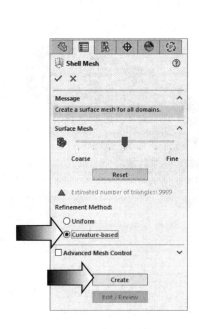

For Refinement Method, select **Curvature-Based**.

Click **Create**.

Click **OK**.

A green check mark on the Shell Mesh node indicates that a shell mesh is created.

After a shell mesh is created, several commands appear under the Shell Mesh node. Right-click on Shell Mesh icon to access these commands.

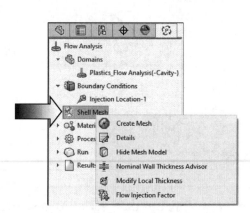

7. Selecting a Polymer:

The default material data base offers thousands of Materials, and the polymers are organized by the family and company. They can be applied to the mold cavities and mold inserts.

From the **PlasticsManager** tree, expand the **Material** option and double-click on **Polymer** to Open Database.

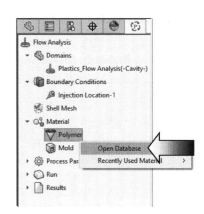

Select the following:
Default Database, Sort by Family, ABS, 22 (P) BASF / ABS6003 (arrows).

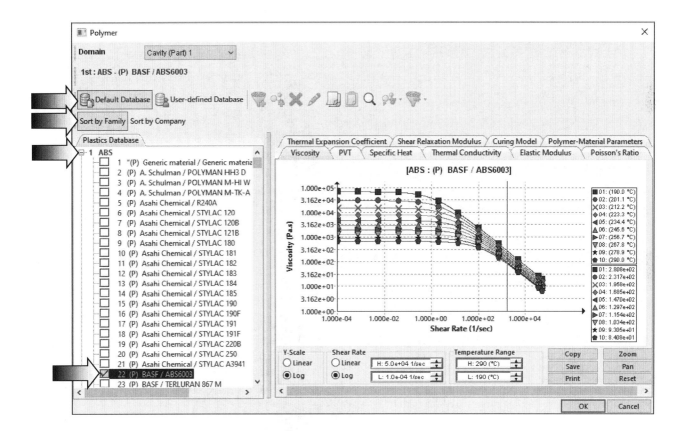

Click **OK**.

Depending on the material selected for the cavity, insert, mold, or cooling channel domain, the complete list of material properties appears in the selected material's dialog box.

At this point, the information needed to run an analysis is completed. The next step is to run the flow analysis and interpret the results.

8. Running the Flow Analysis:

From the **PlasticsManager** tree, expand the **Run** feature. Right-click **Flow** and select: **Run** (arrow).

The Analysis Manager dialog-box appears and the analysis is started. This analysis may take several minutes.

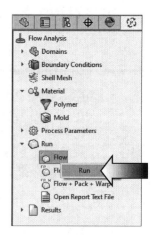

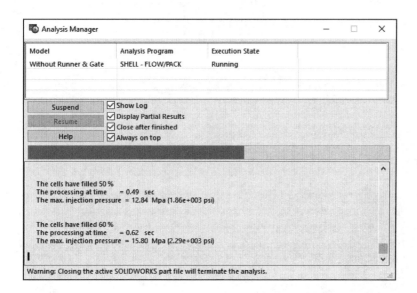

When the analysis is completed, the Flow Results are displayed on the left side of the Plastics-Manager tree. We will take a look at some of them in the next steps.

9. Viewing the Fill Time plot:

Use the **Fill Time** plot to view the profile of the liquid plastic as it flows through the cavity of the mold. The **blue** color regions are the first areas to fill and the **red** are the last one to fill.

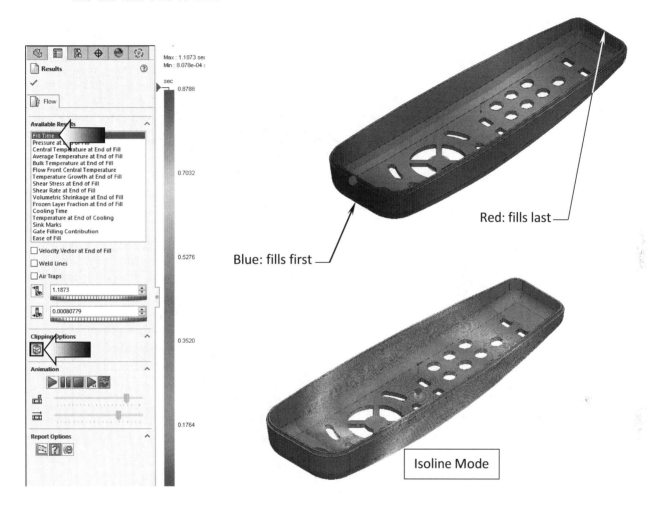

Red: fills last

Blue: fills first

Isoline Mode

The Isoline Mode available for Shell Mesh indicates where particular value is constant.

Under Clipping Options, click the **Isoline** button (arrow).
The Isoline Mode is only available for Shell Mesh.
Isoline plots the regions of plastic material where results fall within the range of min and max values.

The maximum and minimum values of the results shown on the active plot and the **Results Adviser** appears on the right side indicating the part can be successfully filled with an injection pressure of 38.3 MPa (5553.76 psi).

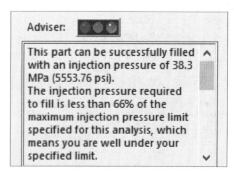

Adviser:

This part can be successfully filled with an injection pressure of 38.3 MPa (5553.76 psi).
The injection pressure required to fill is less than 66% of the maximum injection pressure limit specified for this analysis, which means you are well under your specified limit.

10. Animating the results:

In the same **Results** window, under Animation, click **Play** (arrow).

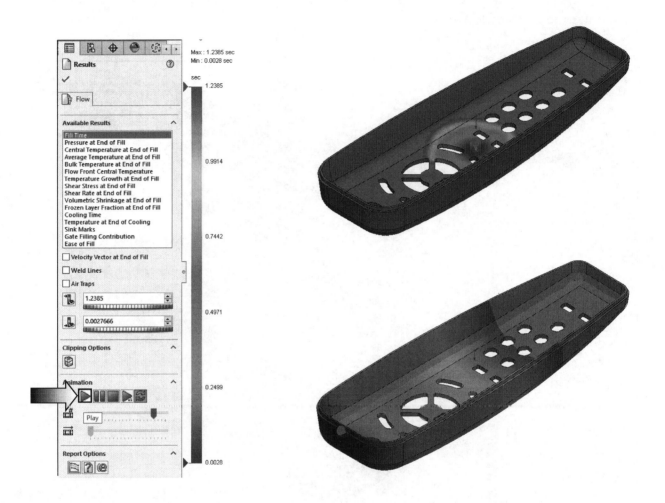

The melt flow front position is animated showing the filling stage. Use the tools in the Animation section to control the animation speed, to pause, stop, or loop the animation.

Starting from the Injection Location, the **blue** color areas get filled first and the **red** color areas are filled last.

Click the **Stop** (square) button to stop the animation but keep the Results dialog box open for the next step.

11. Displaying the Weld Lines:

Weld Lines are formed when two or more flow fronts come together. They appear when there are multiple injection locations or multiple wall thicknesses in the part. To avoid the weld lines, either move the injection locations or make changes in the plastic part, but they cannot be eliminated if there are through holes in the part.

Click **Control + 6** to change to the bottom orientation.

Click the **Weld Lines** checkbox (arrow).

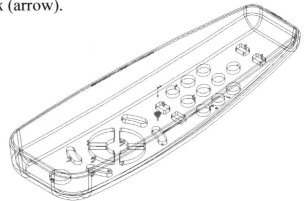

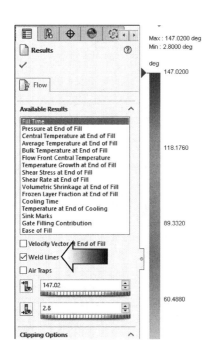

The plot displays the weld line locations and the angle of the flow field as the weld line form.

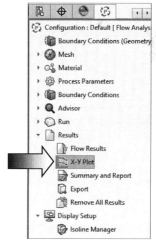

Click **OK** to close the Flow Results.

Double-click the **X-Y Plot** (arrow).
The X-Y plots are used to visualize key results such as the evolution of the Clamp force as a function of the cycle time, or to view pressure buildup at the injection point.

Click **OK** to close the X-Y Plot option.

12. Viewing the results:

Double-click on Flow **Results** to see all other result plots.

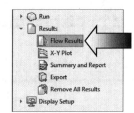

The Fill Time results

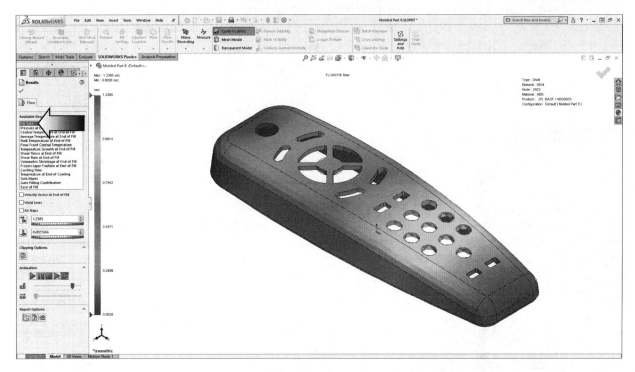

The Pressure at End of Fill results

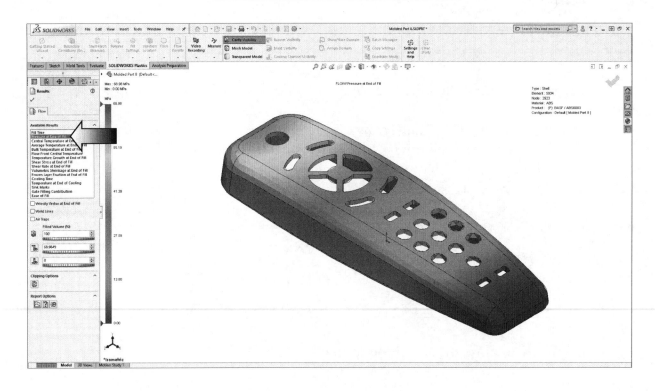

The Central Temperature at End of Fill results

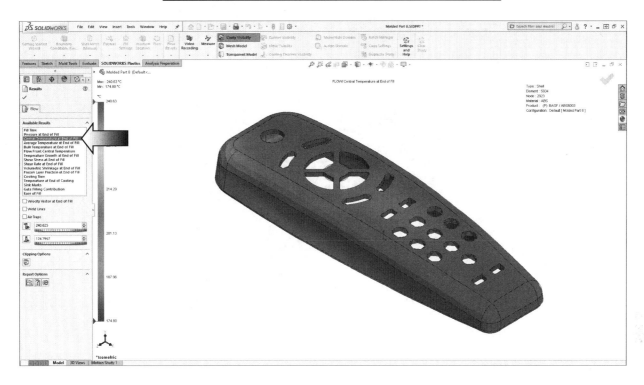

The Average Temperature at End of Fill results

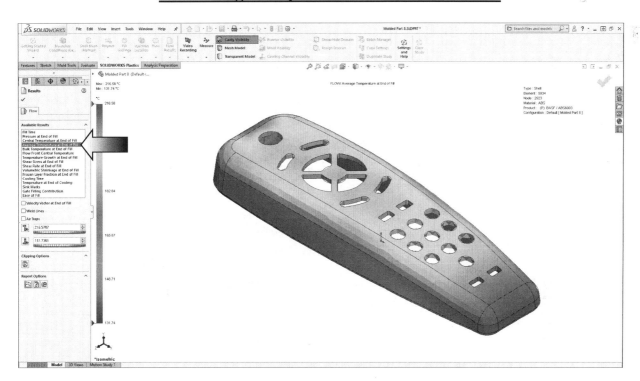

The Bulk Temperature at End of Fill results

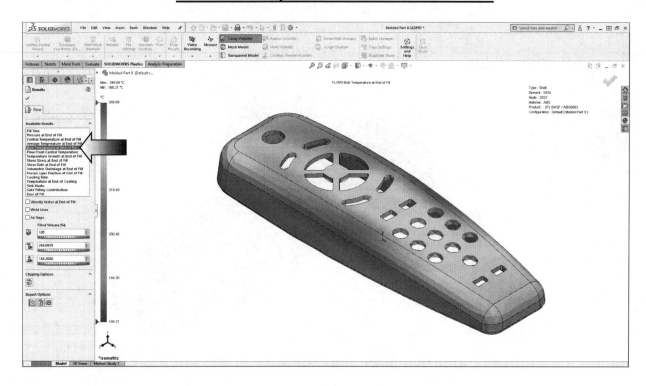

The Flow Front Central Temperature results

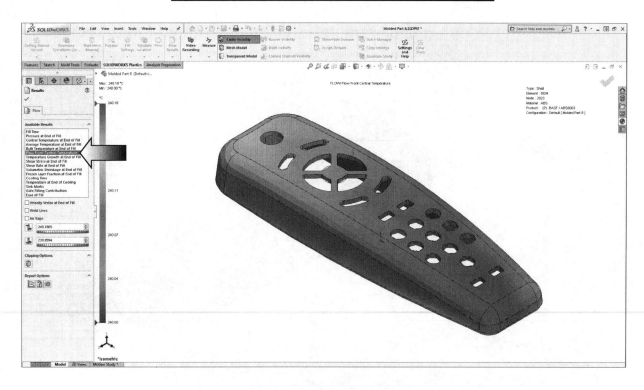

The Temperature Growth at End of Fill results

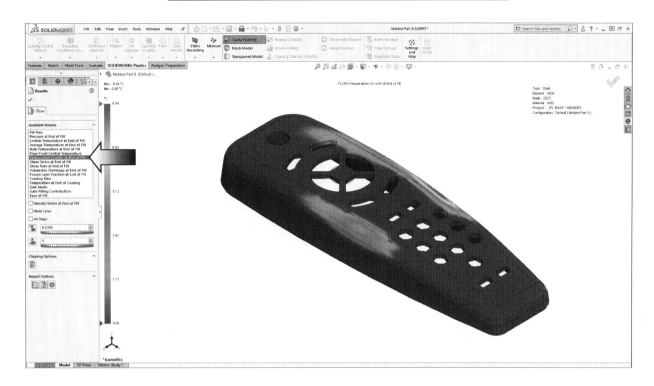

The Shear Stress at End of Fill results

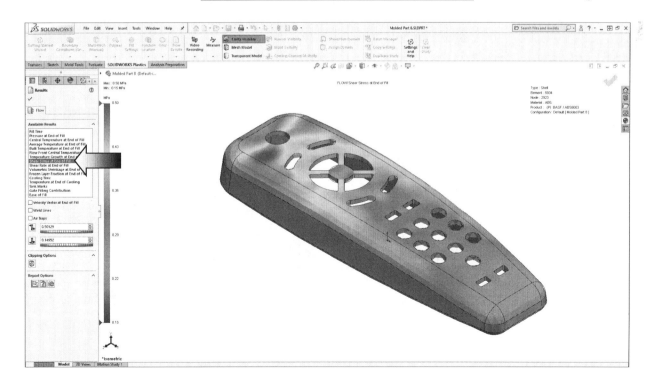

The Shear Rate at End of Fill results

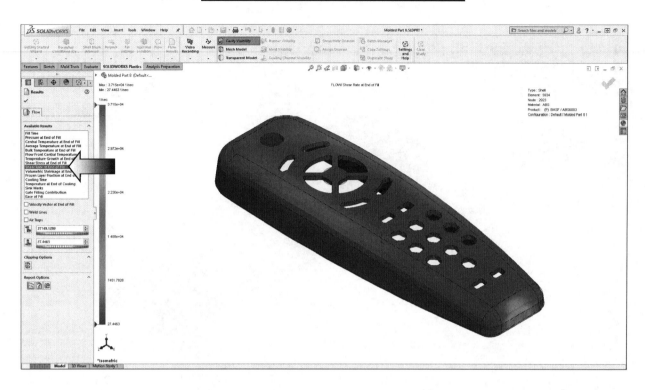

The Volumetric Shrinkage at End of Fill results

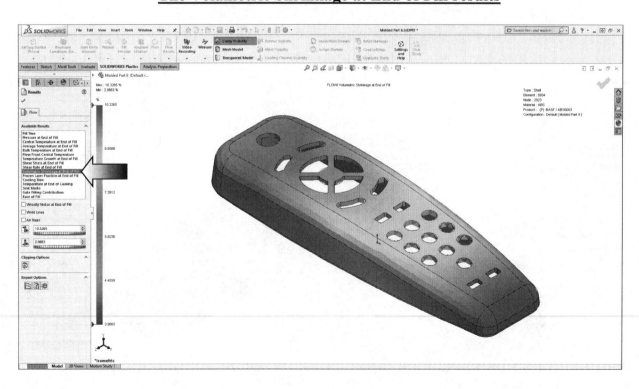

The Frozen Layer Fraction at End of Fill results

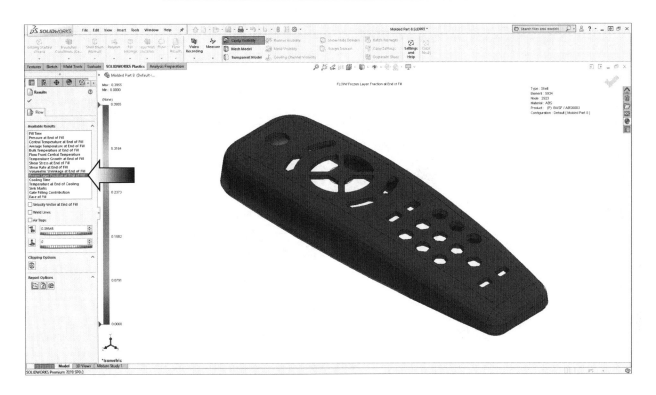

The Cooling Time results

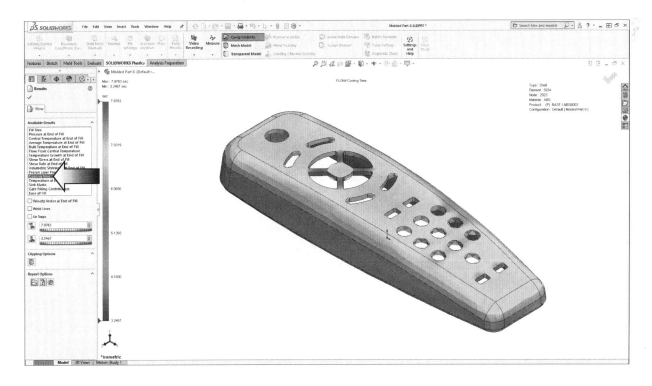

The Temperature at End of Cooling results

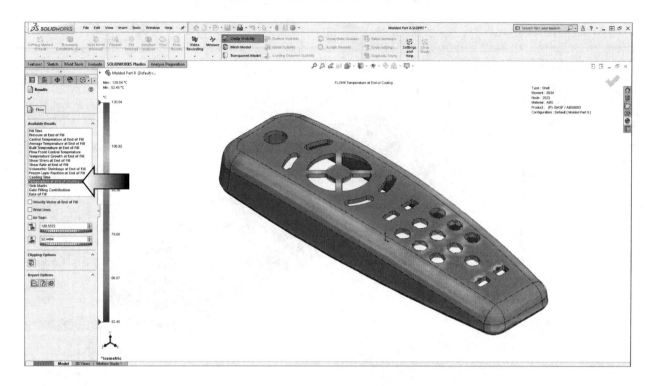

The Sink Marks results

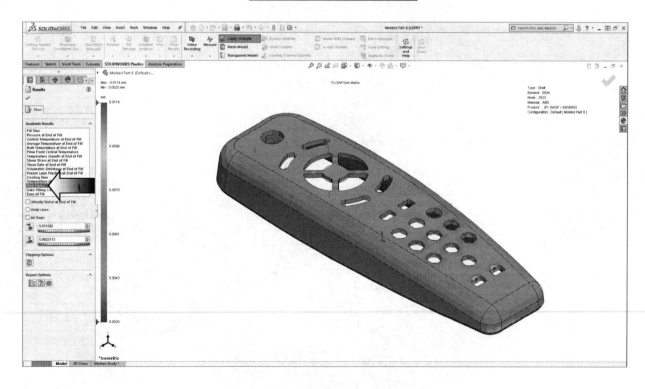

The Gate Filling Distribution results

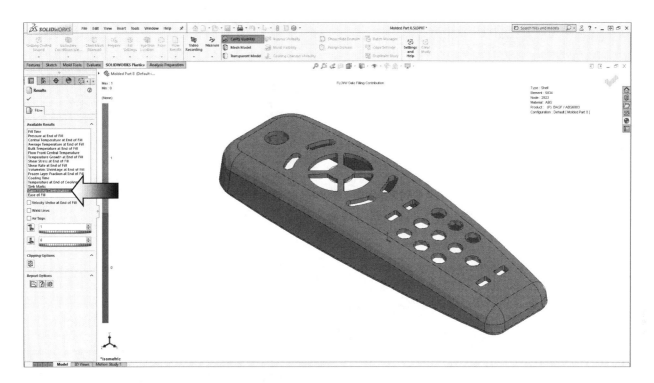

The End of Fill results

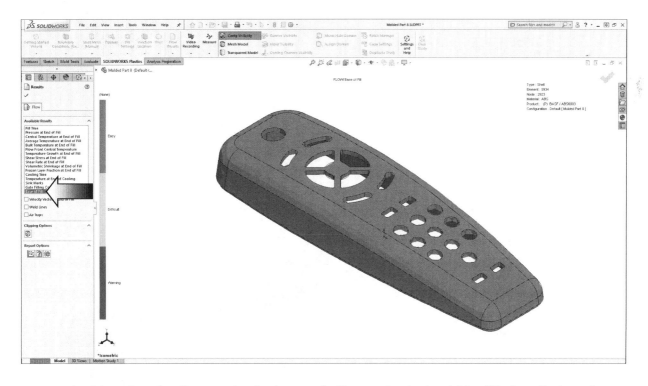

At this point, the Green color in the part indicates that it should be filled easily based on the references such as Material, Injection Location and other parameters provided.

13. Saving your work:

Select **Files, Save As**.

Enter: **Plastics_Flow Analysis_Completed.sldprt** for the file name.

Click **Save**.

Glossary

Alloys:

An Alloy is a mixture of two or more metals (and sometimes a non-metal). The mixture is made by heating and melting the substances together.
Example of alloys are Bronze (Copper and Tin), Brass (Copper and Zinc), and Steel (Iron and Carbon).

Gravity and Mass:

Gravity is the force that pulls everything on earth toward the ground and makes things feel heavy. Gravity makes all falling bodies accelerate at a constant 32ft. per second (9.8 m/s). In the earth's atmosphere, air resistance slows acceleration. Only on airless Moon would a feather and a metal block fall to the ground together.
The mass of an object is the amount of material it contains.
A body with greater mass has more inertia; it needs a greater force to accelerate.
Weight depends on the force of gravity, but mass does not.

When an object spins around another (for example: a satellite orbiting the earth) it is pushed outward. Two forces are at work here: Centrifugal (pushing outward) and Centripetal (pulling inward). If you whirl a ball around you on a string, you pull it inward (Centripetal force). The ball seems to pull outward (Centrifugal force) and if released will fly off in a straight line.

Heat:

Heat is a form of energy and can move from one substance to another in one of three ways: by Convection, by Radiation, and by Conduction.

Convection takes place only in liquids like water (for example: water in a kettle) and gases (for example: air warmed by a heat source such as a fire or radiator).
When liquid or gas is heated, it expands and becomes less dense. Warm air above the radiator rises and cool air moves in to take its place, creating a convection current.

Radiation is movement of heat through the air. Heat forms match set molecules of air moving and rays of heat spread out around the heat source.

Conduction occurs in solids such as metals. The handle of a metal spoon left in boiling liquid warms up as molecules at the heated end move faster and collide with their neighbors, setting them moving. The heat travels through the metal, which is a good conductor of heat.

Inertia:

A body with a large mass is harder to start and also to stop. A heavy truck traveling at 50mph needs more power breaks to stop its motion than a smaller car traveling at the same speed.
Inertia is the tendency of an object either to stay still or to move steadily in a straight line, unless another force (such as a brick wall stopping the vehicle) makes it behave differently.

Joules:

The Joules is the SI unit of work or energy.
One Joule of work is done when a force of one Newton moves through a distance of one meter. The Joule is named after the English scientist James Joule (1818-1889).

Materials:

Stainless steel is an alloy of steel with chromium or nickel.

Steel is made by the basic oxygen process. The raw material is about three parts melted iron and one part scrap steel. Blowing oxygen into the melted iron raises the temperature and gets rid of impurities.

All plastics are chemical compounds called polymers.

Glass is made by mixing and heating sand, limestone, and soda ash. When these ingredients melt they turn into glass, which is hardened when it cools.
Glass is in fact not a solid but a "supercooled" liquid; it can be shaped by blowing, pressing, drawing, casting into molds, rolling, and floating across molten tin, to make large sheets.

Ceramic objects, such as pottery and porcelain, electrical insulators, bricks, and roof tiles are all made from clay. The clay is shaped or molded when wet and soft, and heated in a kiln until it hardens.

Machine Tools:

Are powered tools used for shaping metal or other materials, by drilling holes, chiseling, grinding, pressing, or cutting. Often the material (the work piece) is moved while the tool stays still (lathe), or vice versa, the work piece stays while the tool moves (mill).
Most common machine tools are Mill, Lathe, Saw, Broach, Punch press, Grind, Bore and Stamp break.

CNC

Computer Numerical Control is the automation of machine tools that are operated by precisely programmed commands encoded on a storage medium, as opposed to controlled manually via hand wheels or levers, or mechanically automated via cams alone. Most CNC today is computer numerical control in which computers play an integral part of the control.

3D Printing

All methods work by working in layers, adding material, etc. different to other techniques, which are subtractive. Support is needed because almost all methods could support multi material printing, but it is currently only available in certain top tier machines.

A method of turning digital shapes into physical objects. Due to its nature, it allows us to accurately control the shape of the product. The drawback is size restraints and materials are often not durable.

While FDM does not seem like the best method for instrument manufacturing, it is one of the cheapest and most universally available methods.

EDM
Electric Discharge Machining.

FDM
Fused Deposition Modeling.

SLA
Stereo Lithography.

SLS
Selective Laser Sintering.

SLM
Selective Laser Melting.

J-P
Jetted Photopolymer (or Polyjet).

EDM Electric Discharge Machining

The basic EDM process is really quite simple. An electrical spark is created between an electrode and a work piece. The spark is visible evidence of the flow of electricity. This electric spark produces intense heat with  temperatures reaching 8000 to 12000 degrees Celsius, melting almost anything.

The spark is very carefully controlled and localized so that it only affects the surface of the material.

The EDM process usually does not affect the heat treat below the surface. With wire EDM the spark always takes place in the dielectric of deionized water. The conductivity of the water is carefully controlled making an excellent environment for the EDM process. The water acts as a coolant and flushes away the eroded metal particles.

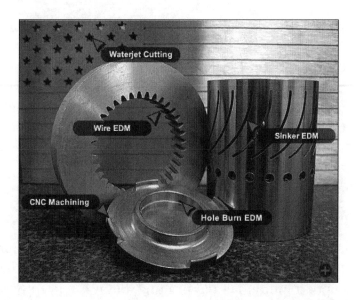

FDM Fused Deposition Modeling

3D printers that run on FDM Technology build parts layer-by-layer by heating thermoplastic material to a semi-liquid state and extruding it according to computer controlled paths.

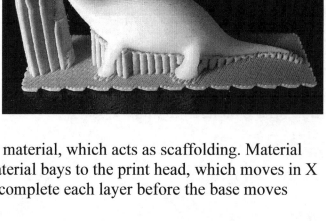

FDM uses two materials to execute a print job: modeling material, which constitutes the finished piece, and support material, which acts as scaffolding. Material filaments are fed from the 3D printer's material bays to the print head, which moves in X and Y coordinates, depositing material to complete each layer before the base moves down the Z axis and the next layer begins.

Once the 3D printer is done building, the user breaks the support material away or dissolves it in detergent and water, and the part is ready to use.

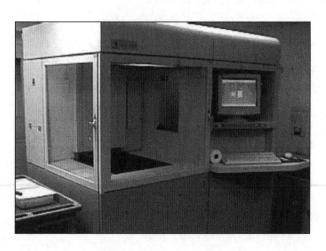

SLA StereoLithograph Apparatus

Stereolithography is an additive fabrication process utilizing a vat of liquid UV-curable photopolymer "resin" and a UV laser to build parts a layer at a time. On each layer, the laser beam traces a part cross-section pattern on the surface of the liquid resin. Exposure to the UV laser light cures, or solidifies the pattern traced on the resin and adheres it to the layer below.

After a pattern has been traced, the SLA's elevator platform descends by a single layer thickness, typically .0019in to .0059in. Then, a resin-filled blade sweeps across the part cross section, re-coating it with fresh material. On this new liquid surface the subsequent layer pattern is traced, adhering to the previous layer.

A complete 3-D part is formed by this process. After building, parts are cleaned of excess resin by immersion in a chemical bath and then cured in a UV oven.

SLS Selective Laser Sintering

Selective laser sintering (SLS) is an additive manufacturing (AM) technique that uses a laser as the power source to sinter (compacting) metal, aiming the laser automatically at points in space defined by a 3D model, binding the material together to create a solid structure.

It is similar to direct metal laser sintering (DMLS); the two are instantiations of the same concept but differ in technical details. Selective laser melting (SLM) uses a comparable concept, but in SLM the material is fully melted rather than sintered,[1] allowing different properties (crystal structure, porosity, and so on). SLS (as well as the other mentioned AM techniques) is a relatively new technology that so far has mainly been used for rapid prototyping and for low-volume production of component parts. Production roles are expanding as the commercialization of AM technology improves.

SLM Selective Laser Melting

Selective laser melting is an additive manufacturing process that uses 3D CAD data as a digital information source and energy in the form of a high-power laser beam, to create three-dimensional metal parts by fusing fine metal powders together. Manufacturing applications in aerospace or medical orthopedics are being pioneered.

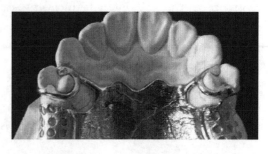

The process starts by slicing the 3D CAD file data into layers, usually from 20 to 100 micrometres thick (0.00078740157 to 0.00393700787 in) creating a 2D image of each layer; this file format is the industry standard .stl file used in most layer-based 3D printing or stereolithography technologies.

This file is then loaded into a file preparation software package that assigns parameters, values and physical supports that allow the file to be interpreted and built by different types of additive manufacturing machines.

J-P Jetted Photopolymer (or Polyjet)

Photopolymer jetting (or PolyJet) builds prototypes by jetting liquid photopolymer resin from ink-jet style heads. The resin is sprayed from the moving heads, and only the amount of material needed is used.

UV light is simultaneously emitted from the head, which cures each layer of resin immediately after it is applied. The process produces excellent surface finish and feature detail. Photopolymer jetting is used primarily to check form and fit and can handle limited functional tests due to the limited strength of photopolymer resins.

This process offers the unique ability to create prototypes with more than one type of material. For instance, a toothbrush prototype could be composed with a rigid shaft with a rubber-like over-molding for grip.

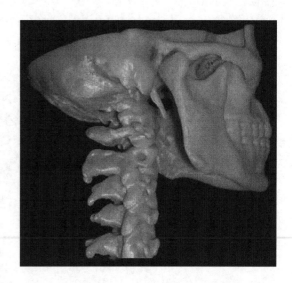

The process works with a variety of proprietary photopolymer resins as opposed to production materials. A tradeoff with this technology is that exposure to ambient heat, humidity, or sunlight can cause dimensional change that can affect tolerance. The process is faster and cleaner than the traditional vat and laser photo-polymer processes.

Carbon 3D

The Carbon 3D not only prints composite materials like carbon fiber, but also fiberglass, nylon and PLA. Of course, only one at a time.

The printer employs some pretty nifty advancements, too, including a self-leveling printing bed that clicks into position before each print.

The Carbon 3D is groundbreaking 3D printing technology which is 25 to 100 times faster than currently available commercial PolyJet or SLA machines.

It is a true quantum leap forward for 3D printing speed!

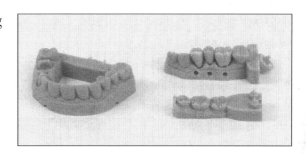

Newton's Law:

1. Every object remains stopped or goes on moving at a steady rate in a straight line unless acted upon by another force. This is the inertia principle.
2. The amount of force needed to make an object change its speed depends on the mass of the object and the amount of the acceleration or deceleration required.
3. To every action there is an equal and opposite reaction. When a body is pushed on way by a force, another force pushes back with equal strength.

Polymers:

A polymer is made of one or more large molecules formed from thousands of smaller molecules. Rubber and Wood are natural polymers. Plastics are synthetic (artificially made) polymers.

Speed and Velocity:

Speed is the rate at which a moving object changes position (how far it moves in a fixed time).
Velocity is speed in a particular direction.
If either speed or direction is changed, velocity also changes.

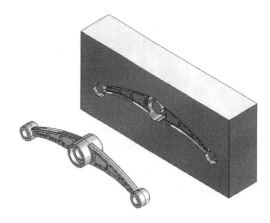

Absorbed
A feature, sketch, or annotation that is contained in another item (usually a feature) in the FeatureManager design tree. Examples are the profile sketch and profile path in a base-sweep, or a cosmetic thread annotation in a hole.

Align
Tools that assist in lining up annotations and dimensions (left, right, top, bottom, and so on). For aligning parts in an assembly.

Alternate position view
A drawing view in which one or more views are superimposed in phantom lines on the original view. Alternate position views are often used to show range of motion of an assembly.

Anchor point
The end of a leader that attaches to the note, block, or other annotation. Sheet formats contain anchor points for a bill of materials, a hole table, a revision table, and a weldment cut list.

Annotation
A text note or a symbol that adds specific design intent to a part, assembly, or drawing. Specific types of annotations include note, hole callout, surface finish symbol, datum feature symbol, datum target, geometric tolerance symbol, weld symbol, balloon, and stacked balloon. Annotations that apply only to drawings include center mark, annotation centerline, area hatch, and block.

Appearance callouts
Callouts that display the colors and textures of the face, feature, body, and part under the entity selected and are a shortcut to editing colors and textures.

Area hatch
A crosshatch pattern or fill applied to a selected face or to a closed sketch in a drawing.

Assembly
A document in which parts, features, and other assemblies (sub-assemblies) are mated together. The parts and sub-assemblies exist in documents separate from the assembly. For example, in an assembly, a piston can be mated to other parts, such as a connecting rod or cylinder. This new assembly can then be used as a sub-assembly in an assembly of an engine. The extension for a SOLIDWORKS assembly file name is .SLDASM.

Attachment point
The end of a leader that attaches to the model (to an edge, vertex, or face, for example) or to a drawing sheet.

Axis

A straight line that can be used to create model geometry, features, or patterns. An axis can be made in a number of different ways, including using the intersection of two planes.

Balloon

Labels parts in an assembly, typically including item numbers and quantity. In drawings, the item numbers are related to rows in a bill of materials.

Base

The first solid feature of a part.

Baseline dimensions

Sets of dimensions measured from the same edge or vertex in a drawing.

Bend

A feature in a sheet metal part. A bend generated from a filleted corner, cylindrical face, or conical face is a round bend; a bend generated from sketched straight lines is a sharp bend.

Bill of materials

A table inserted into a drawing to keep a record of the parts used in an assembly.

Block

A user-defined annotation that you can use in parts, assemblies, and drawings. A block can contain text, sketch entities (except points), and area hatch, and it can be saved in a file for later use as, for example, a custom callout or a company logo.

Bottom-up assembly

An assembly modeling technique where you create parts and then insert them into an assembly.

Broken-out section

A drawing view that exposes inner details of a drawing view by removing material from a closed profile, usually a spline.

Cavity

The mold half that holds the cavity feature of the design part.

Center mark

A cross that marks the center of a circle or arc.

Centerline

A centerline marks, in phantom font, an axis of symmetry in a sketch or drawing.

Chamfer

Bevels a selected edge or vertex. You can apply chamfers to both sketches and features.

Child

A dependent feature related to a previously built feature. For example, a chamfer on the edge of a hole is a child of the parent hole.

Click-release

As you sketch, if you click and then release the pointer, you are in click-release mode. Move the pointer and click again to define the next point in the sketch sequence.

Click-drag

As you sketch, if you click and drag the pointer, you are in click-drag mode. When you release the pointer, the sketch entity is complete.

Closed profile

Also called a closed contour, it is a sketch or sketch entity with no exposed endpoints, for example, a circle or polygon.

Collapse

The opposite of explode. The collapse action returns an exploded assembly's parts to their normal positions.

Collision Detection

An assembly function that detects collisions between components when components move or rotate. A collision occurs when an entity on one component coincides with any entity on another component.

Component

Any part or sub-assembly within an assembly.

Configuration

A variation of a part or assembly within a single document. Variations can include different dimensions, features, and properties. For example, a single part such as a bolt can contain different configurations that vary the diameter and length.

ConfigurationManager

Located on the left side of the SOLIDWORKS window, it is a means to create, select, and view the configurations of parts and assemblies.

Constraint

The relations between sketch entities, or between sketch entities and planes, axes, edges, or vertices.

Construction geometry
The characteristic of a sketch entity that the entity is used in creating other geometry but is not itself used in creating features.

Coordinate system
A system of planes used to assign Cartesian coordinates to features, parts, and assemblies. Part and assembly documents contain default coordinate systems; other coordinate systems can be defined with reference geometry. Coordinate systems can be used with measurement tools and for exporting documents to other file formats.

Cosmetic thread
An annotation that represents threads.

Crosshatch
A pattern (or fill) applied to drawing views such as section views and broken-out sections.

Curvature
Curvature is equal to the inverse of the radius of the curve. The curvature can be displayed in different colors according to the local radius (usually of a surface).

Cut
A feature that removes material from a part by such actions as extrude, revolve, loft, sweep, thicken, cavity, and so on.

Dangling
A dimension, relation, or drawing section view that is unresolved. For example, if a piece of geometry is dimensioned, and that geometry is later deleted, the dimension becomes dangling.

Degrees of freedom
Geometry that is not defined by dimensions or relations is free to move. In 2D sketches, there are three degrees of freedom: movement along the X and Y axes, and rotation about the Z axis (the axis normal to the sketch plane). In 3D sketches and in assemblies, there are six degrees of freedom: movement along the X, Y, and Z axes, and rotation about the X, Y, and Z axes.

Derived part
A derived part is a new base, mirror, or component part created directly from an existing part and linked to the original part such that changes to the original part are reflected in the derived part.

Derived sketch

A copy of a sketch, in either the same part or the same assembly that is connected to the original sketch. Changes in the original sketch are reflected in the derived sketch.

Design Library

Located in the Task Pane, the Design Library provides a central location for reusable elements such as parts, assemblies, and so on.

Design table

An Excel spreadsheet that is used to create multiple configurations in a part or assembly document.

Detached drawing

A drawing format that allows opening and working in a drawing without loading the corresponding models into memory. The models are loaded on an as-needed basis.

Detail view

A portion of a larger view, usually at a larger scale than the original view.

Dimension line

A linear dimension line references the dimension text to extension lines indicating the entity being measured. An angular dimension line references the dimension text directly to the measured object.

DimXpertManager

Located on the left side of the SOLIDWORKS window, it is a means to manage dimensions and tolerances created using DimXpert for parts according to the requirements of the ASME Y.14.41-2003 standard.

DisplayManager

The DisplayManager lists the appearances, decals, lights, scene, and cameras applied to the current model. From the DisplayManager, you can view applied content, and add, edit, or delete items. When PhotoView 360 is added in, the DisplayManager also provides access to PhotoView options.

Document

A file containing a part, assembly, or drawing.

Draft

The degree of taper or angle of a face usually applied to molds or castings.

Drawing

A 2D representation of a 3D part or assembly. The extension for a SOLIDWORKS drawing file name is .SLDDRW.

Drawing sheet

A page in a drawing document.

Driven dimension

Measurements of the model, but they do not drive the model and their values cannot be changed.

Driving dimension

Also referred to as a model dimension, it sets the value for a sketch entity. It can also control distance, thickness, and feature parameters.

Edge

A single outside boundary of a feature.

Edge flange

A sheet metal feature that combines a bend and a tab in a single operation.

Equation

Creates a mathematical relation between sketch dimensions, using dimension names as variables, or between feature parameters, such as the depth of an extruded feature or the instance count in a pattern.

Exploded view

Shows an assembly with its components separated from one another, usually to show how to assemble the mechanism.

Export

Save a SOLIDWORKS document in another format for use in other CAD/CAM, rapid prototyping, web, or graphics software applications.

Extension line

The line extending from the model indicating the point from which a dimension is measured.

Extrude

A feature that linearly projects a sketch to either add material to a part (in a base or boss) or remove material from a part (in a cut or hole).

Face

A selectable area (planar or otherwise) of a model or surface with boundaries that help define the shape of the model or surface. For example, a rectangular solid has six faces.

Fasteners
A SOLIDWORKS Toolbox library that adds fasteners automatically to holes in an assembly.

Feature
An individual shape that, combined with other features, makes up a part or assembly. Some features, such as bosses and cuts, originate as sketches. Other features, such as shells and fillets, modify a feature's geometry. However, not all features have associated geometry. Features are always listed in the FeatureManager design tree.

FeatureManager design tree
Located on the left side of the SOLIDWORKS window, it provides an outline view of the active part, assembly, or drawing.

Fill
A solid area hatch or crosshatch. Fill also applies to patches on surfaces.

Fillet
An internal rounding of a corner or edge in a sketch, or an edge on a surface or solid.

Forming tool
Dies that bend, stretch, or otherwise form sheet metal to create such form features as louvers, lances, flanges, and ribs.

Fully defined
A sketch where all lines and curves in the sketch, and their positions, are described by dimensions or relations, or both, and cannot be moved. Fully defined sketch entities are shown in black.

Geometric tolerance
A set of standard symbols that specify the geometric characteristics and dimensional requirements of a feature.

Graphics area
The area in the SOLIDWORKS window where the part, assembly, or drawing appears.

Guide curve
A 2D or 3D curve used to guide a sweep or loft.

Handle
An arrow, square, or circle that you can drag to adjust the size or position of an entity (a feature, dimension, or sketch entity, for example).

Helix

A curve defined by pitch, revolutions, and height. A helix can be used, for example, as a path for a swept feature cutting threads in a bolt.

Hem

A sheet metal feature that folds back at the edge of a part. A hem can be open, closed, double, or teardrop.

HLR

(Hidden lines removed) a view mode in which all edges of the model that are not visible from the current view angle are removed from the display.

HLV

(Hidden lines visible) A view mode in which all edges of the model that are not visible from the current view angle are shown gray or dashed.

Import

Open files from other CAD software applications into a SOLIDWORKS document.

In-context feature

A feature with an external reference to the geometry of another component; the in-context feature changes automatically if the geometry of the referenced model or feature changes.

Inference

The system automatically creates (infers) relations between dragged entities (sketched entities, annotations, and components) and other entities and geometry. This is useful when positioning entities relative to one another.

Instance

An item in a pattern or a component in an assembly that occurs more than once. Blocks are inserted into drawings as instances of block definitions.

Interference detection

A tool that displays any interference between selected components in an assembly.

Jog

A sheet metal feature that adds material to a part by creating two bends from a sketched line.

Knit

A tool that combines two or more faces or surfaces into one. The edges of the surfaces must be adjacent and not overlapping, but they cannot ever be planar. There is no difference in the appearance of the face or the surface after knitting.

Layout sketch
A sketch that contains important sketch entities, dimensions, and relations. You reference the entities in the layout sketch when creating new sketches, building new geometry, or positioning components in an assembly. This allows for easier updating of your model because changes you make to the layout sketch propagate to the entire model.

Leader
A solid line from an annotation (note, dimension, and so on) to the referenced feature.

Library feature
A frequently used feature, or combination of features, that is created once and then saved for future use.

Lightweight
A part in an assembly or a drawing has only a subset of its model data loaded into memory. The remaining model data is loaded on an as-needed basis. This improves performance of large and complex assemblies.

Line
A straight sketch entity with two endpoints. A line can be created by projecting an external entity such as an edge, plane, axis, or sketch curve into the sketch.

Loft
A base, boss, cut, or surface feature created by transitions between profiles.

Lofted bend
A sheet metal feature that produces a roll form or a transitional shape from two open profile sketches. Lofted bends often create funnels and chutes.

Mass properties
A tool that evaluates the characteristics of a part or an assembly such as volume, surface area, centroid, and so on.

Mate
A geometric relationship, such as coincident, perpendicular, tangent, and so on, between parts in an assembly.

Mate reference
Specifies one or more entities of a component to use for automatic mating. When you drag a component with a mate reference into an assembly, the software tries to find other combinations of the same mate reference name and mate type.

Mates folder

A collection of mates that are solved together. The order in which the mates appear within the Mates folder does not matter.

Mirror

(a) A mirror feature is a copy of a selected feature, mirrored about a plane or planar face.
(b) A mirror sketch entity is a copy of a selected sketch entity that is mirrored about a centerline.

Miter flange

A sheet metal feature that joins multiple edge flanges together and miters the corner.

Model

3D solid geometry in a part or assembly document. If a part or assembly document contains multiple configurations, each configuration is a separate model.

Model dimension

A dimension specified in a sketch or a feature in a part or assembly document that defines some entity in a 3D model.

Model item

A characteristic or dimension of feature geometry that can be used in detailing drawings.

Model view

A drawing view of a part or assembly.

Mold

A set of manufacturing tooling used to shape molten plastic or other material into a designed part. You design the mold using a sequence of integrated tools that result in cavity and core blocks that are derived parts of the part to be molded.

Motion Study

Motion Studies are graphical simulations of motion and visual properties with assembly models. Analogous to a configuration, they do not actually change the original assembly model or its properties. They display the model as it changes based on simulation elements you add.

Multibody part

A part with separate solid bodies within the same part document. Unlike the components in an assembly, multibody parts are not dynamic.

Native format
DXF and DWG files remain in their original format (are not converted into SOLIDWORKS format) when viewed in SOLIDWORKS drawing sheets (view only).

Open profile
Also called an open contour, it is a sketch or sketch entity with endpoints exposed. For example, a U-shaped profile is open.

Ordinate dimensions
A chain of dimensions measured from a zero ordinate in a drawing or sketch.

Origin
The model origin appears as three gray arrows and represents the (0,0,0) coordinate of the model. When a sketch is active, a sketch origin appears in red and represents the (0,0,0) coordinate of the sketch. Dimensions and relations can be added to the model origin but not to a sketch origin.

Out-of-context feature
A feature with an external reference to the geometry of another component that is not open.

Over defined
A sketch is over defined when dimensions or relations are either in conflict or redundant.

Parameter
A value used to define a sketch or feature (often a dimension).

Parent
An existing feature upon which other features depend. For example, in a block with a hole, the block is the parent to the child hole feature.

Part
A single 3D object made up of features. A part can become a component in an assembly, and it can be represented in 2D in a drawing. Examples of parts are bolt, pin, plate, and so on. The extension for a SOLIDWORKS part file name is .SLDPRT.

Path
A sketch, edge, or curve used in creating a sweep or loft.

Pattern
A pattern repeats selected sketch entities, features, or components in an array, which can be linear, circular, or sketch driven. If the seed entity is changed, the other instances in

the pattern update.

Physical Dynamics

An assembly tool that displays the motion of assembly components in a realistic way. When you drag a component, the component applies a force to other components it touches. Components move only within their degrees of freedom.

Pierce relation

Makes a sketch point coincident to the location at which an axis, edge, line, or spline pierces the sketch plane.

Planar

Entities that can lie on one plane. For example, a circle is planar, but a helix is not.

Plane

Flat construction geometry. Planes can be used for a 2D sketch, section view of a model, a neutral plane in a draft feature, and others.

Point

A singular location in a sketch, or a projection into a sketch at a single location of an external entity (origin, vertex, axis, or point in an external sketch).

Predefined view

A drawing view in which the view position, orientation, and so on can be specified before a model is inserted. You can save drawing documents with predefined views as templates.

Profile

A sketch entity used to create a feature (such as a loft) or a drawing view (such as a detail view). A profile can be open (such as a U shape or open spline) or closed (such as a circle or closed spline).

Projected dimension

If you dimension entities in an isometric view, projected dimensions are the flat dimensions in 2D.

Projected view

A drawing view projected orthogonally from an existing view.

PropertyManager

Located on the left side of the SOLIDWORKS window, it is used for dynamic editing of sketch entities and most features.

RealView graphics

A hardware (graphics card) support of advanced shading in real time; the rendering applies to the model and is retained as you move or rotate a part.

Rebuild

Tool that updates (or regenerates) the document with any changes made since the last time the model was rebuilt. Rebuild is typically used after changing a model dimension.

Reference dimension

A dimension in a drawing that shows the measurement of an item but cannot drive the model and its value cannot be modified. When model dimensions change, reference dimensions update.

Reference geometry

Includes planes, axes, coordinate systems, and 3D curves. Reference geometry is used to assist in creating features such as lofts, sweeps, drafts, chamfers, and patterns.

Relation

A geometric constraint between sketch entities or between a sketch entity and a plane, axis, edge, or vertex. Relations can be added automatically or manually.

Relative view

A relative (or relative to model) drawing view is created relative to planar surfaces in a part or assembly.

Reload

Refreshes shared documents. For example, if you open a part file for read-only access while another user makes changes to the same part, you can reload the new version, including the changes.

Reorder

Reordering (changing the order of) items is possible in the FeatureManager design tree. In parts, you can change the order in which features are solved. In assemblies, you can control the order in which components appear in a bill of materials.

Replace

Substitutes one or more open instances of a component in an assembly with a different component.

Resolved

A state of an assembly component (in an assembly or drawing document) in which it is fully loaded in memory. All the component's model data is available, so its entities can be selected, referenced, edited, and used in mates, and so on.

Revolve

A feature that creates a base or boss, a revolved cut, or revolved surface by revolving one or more sketched profiles around a centerline.

Rip

A sheet metal feature that removes material at an edge to allow a bend.

Rollback

Suppresses all items below the rollback bar.

Section

Another term for profile in sweeps.

Section line

A line or centerline sketched in a drawing view to create a section view.

Section scope

Specifies the components to be left uncut when you create an assembly drawing section view.

Section view

A section view (or section cut) is (1) a part or assembly view cut by a plane, or (2) a drawing view created by cutting another drawing view with a section line.

Seed

A sketch or an entity (a feature, face, or body) that is the basis for a pattern. If you edit the seed, the other entities in the pattern are updated.

Shaded

Displays a model as a colored solid.

Shared values

Also called linked values, these are named variables that you assign to set the value of two or more dimensions to be equal.

Sheet format

Includes page size and orientation, standard text, borders, title blocks, and so on. Sheet formats can be customized and saved for future use. Each sheet of a drawing document can have a different format.

Shell

A feature that hollows out a part, leaving open the selected faces and thin walls on the remaining faces. A hollow part is created when no faces are selected to be open.

Sketch

A collection of lines and other 2D objects on a plane or face that forms the basis for a feature such as a base or a boss. A 3D sketch is non-planar and can be used to guide a sweep or loft, for example.

Smart Fasteners

Automatically adds fasteners (bolts and screws) to an assembly using the SOLIDWORKS Toolbox library of fasteners.

SmartMates

An assembly mating relation that is created automatically.

Solid sweep

A cut sweep created by moving a tool body along a path to cut out 3D material from a model.

Spiral

A flat or 2D helix, defined by a circle, pitch, and number of revolutions.

Spline

A sketched 2D or 3D curve defined by a set of control points.

Split line

Projects a sketched curve onto a selected model face, dividing the face into multiple faces so that each can be selected individually. A split line can be used to create draft features, to create face blend fillets, and to radiate surfaces to cut molds.

Stacked balloon

A set of balloons with only one leader. The balloons can be stacked vertically (up or down) or horizontally (left or right).

Standard 3 views

The three orthographic views (front, right, and top) that are often the basis of a drawing.

StereoLithography

The process of creating rapid prototype parts using a faceted mesh representation in STL files.

Sub-assembly

An assembly document that is part of a larger assembly. For example, the steering mechanism of a car is a sub-assembly of the car.

Suppress

Removes an entity from the display and from any calculations in which it is involved. You can suppress features, assembly components, and so on. Suppressing an entity does not delete the entity; you can un-suppress the entity to restore it.

Surface

A zero-thickness planar or 3D entity with edge boundaries. Surfaces are often used to create solid features. Reference surfaces can be used to modify solid features.

Sweep

Creates a base, boss, cut, or surface feature by moving a profile (section) along a path. For cut sweeps, you can create solid sweeps by moving a tool body along a path.

Tangent arc

An arc that is tangent to another entity, such as a line.

Tangent edge

The transition edge between rounded or filleted faces in hidden lines visible or hidden lines removed modes in drawings.

Task Pane

Located on the right-side of the SOLIDWORKS window, the Task Pane contains SOLIDWORKS Resources, the Design Library, and the File Explorer.

Template

A document (part, assembly, or drawing) that forms the basis of a new document. It can include user-defined parameters, annotations, predefined views, geometry, and so on.

Temporary axis

An axis created implicitly for every conical or cylindrical face in a model.

Thin feature

An extruded or revolved feature with constant wall thickness. Sheet metal parts are typically created from thin features.

TolAnalyst

A tolerance analysis application that determines the effects that dimensions and tolerances have on parts and assemblies.

Top-down design

An assembly modeling technique where you create parts in the context of an assembly by referencing the geometry of other components. Changes to the referenced components propagate to the parts that you create in context.

Triad

Three axes with arrows defining the X, Y, and Z directions. A reference triad appears in part and assembly documents to assist in orienting the viewing of models. Triads also assist when moving or rotating components in assemblies.

Under defined

A sketch is under defined when there are not enough dimensions and relations to prevent entities from moving or changing size.

Vertex

A point at which two or more lines or edges intersect. Vertices can be selected for sketching, dimensioning, and many other operations.

Viewports

Windows that display views of models. You can specify one, two, or four viewports. Viewports with orthogonal views can be linked, which links orientation and rotation.

Virtual sharp

A sketch point at the intersection of two entities after the intersection itself has been removed by a feature such as a fillet or chamfer. Dimensions and relations to the virtual sharp are retained even though the actual intersection no longer exists.

Weldment

A multibody part with structural members.

Weldment cut list

A table that tabulates the bodies in a weldment along with descriptions and lengths.

Wireframe

A view mode in which all edges of the part or assembly are displayed.

Zebra stripes

Simulate the reflection of long strips of light on a very shiny surface. They allow you to see small changes in a surface that may be hard to see with a standard display.

Zoom

To simulate movement toward or away from a part or an assembly.

Index

SOLIDWORKS Quick Guide

STANDARD Toolbar

 Creates a new document.

 Opens an existing document.

 Saves an active document.

 Make Drawing from Part/Assembly.

 Make Assembly from Part/Assembly.

 Prints the active document.

 Print preview.

 Cuts the selection & puts it on the clipboard.

 Copies the selection & puts it on the clipboard.

 Inserts the clipboard contents.

 Deletes the selection.

 Reverses the last action.

 Rebuilds the part / assembly / drawing.

 Redo the last action that was undone.

 Saves all documents.

 Edits material.

 Closes an existing document.

 Shows or hides the Selection Filter toolbar.

 Shows or hides the Web toolbar.

 Properties.

 File properties.

 Loads or unloads the 3D instant website add-in.

 Select tool.

 Select the entire document.

 Checks read-only files.

 Options.

 Help.

 Full screen view.

 OK.

 Cancel.

 Magnified selection.

SKETCH TOOLS Toolbar

 Select.

 Sketch.

 3D Sketch.

 Sketches a rectangle from the center.

 Sketches a centerpoint arc slot.

 Sketches a 3-point arc slot.

 Sketches a straight slot.

 Sketches a centerpoint straight slot.

 Sketches a 3-point arc.

 Creates sketched ellipses.

Quick Reference Guide to SOLIDWORKS Command Icons & Toolbars

SKETCH TOOLS Toolbar

3D sketch on plane.	Partial ellipses.
Sets up Grid parameters.	Adds a Parabola.
Creates a sketch on a selected plane or face.	Adds a spline.
Equation driven curve.	Sketches a polygon.
Modifies a sketch.	Sketches a corner rectangle.
Copies sketch entities.	Sketches a parallelogram.
Scales sketch entities.	Creates points.
Rotates sketch entities.	Creates sketched centerlines.
Sketches 3 point rectangle from the center.	Adds text to sketch.
Sketches 3 point corner rectangle.	Converts selected model edges or sketch entities to sketch segments.
Sketches a line.	Creates a sketch along the intersection of multiple bodies.
Creates a center point arc: center, start, end.	Converts face curves on the selected face into 3D sketch entities.
Creates an arc tangent to a line.	Mirrors selected segments about a centerline.
Sketches splines on a surface or face.	Fillets the corner of two lines.
Sketches a circle.	Creates a chamfer between two sketch entities.
Sketches a circle by its perimeter.	Creates a sketch curve by offsetting model edges or sketch entities at a specified distance.
Makes a path of sketch entities.	Trims a sketch segment.
Mirrors entities dynamically about a centerline.	Extends a sketch segment.
Insert a plane into the 3D sketch.	Splits a sketch segment.
Instant 2D.	Construction Geometry.
Sketch numeric input.	Creates linear steps and repeat of sketch entities.
Detaches segment on drag.	Creates circular steps and repeat of sketch entities.
Sketch picture.	

Quick Reference Guide to SOLIDWORKS Command Icons & Toolbars

SHEET METAL Toolbar

 Add a bend from a selected sketch in a Sheet Metal part.

 Shows flat pattern for this sheet metal part.

 Shows part without inserting any bends.

 Inserts a rip feature to a sheet metal part.

 Create a Sheet Metal part or add material to existing Sheet Metal part.

 Inserts a Sheet Metal Miter Flange feature.

 Folds selected bends.

 Unfolds selected bends.

 Inserts bends using a sketch line.

 Inserts a flange by pulling an edge.

 Inserts a sheet metal corner feature.

 Inserts a Hem feature by selecting edges.

 Breaks a corner by filleting/chamfering it.

 Inserts a Jog feature using a sketch line.

 Inserts a lofted bend feature using 2 sketches.

 Creates inverse dent on a sheet metal part.

 Trims out material from a corner, in a sheet metal part.

 Inserts a fillet weld bead.

 Converts a solid/surface into a sheet metal part.

 Adds a Cross Break feature into a selected face.

 Sweeps an open profile along an open/closed path.

 Adds a gusset/rib across a bend.

 Corner relief.

 Welds the selected corner.

SURFACES Toolbar

 Creates mid surfaces between offset face pairs.

 Patches surface holes and external edges.

 Creates an extruded surface.

 Creates a revolved surface.

 Creates a swept surface.

 Creates a lofted surface.

 Creates an offset surface.

 Radiates a surface originating from a curve, parallel to a plane.

 Knits surfaces together.

 Creates a planar surface from a sketch or a set of edges.

 Creates a surface by importing data from a file.

 Extends a surface.

 Trims a surface.

 Surface flatten.

 Deletes Face(s).

 Replaces Face with Surface.

 Patches surface holes and external edges by extending the surfaces.

 Creates parting surfaces between core & cavity surfaces.

 Inserts ruled surfaces from edges.

WELDMENTS Toolbar

 Creates a weldment feature.

 Creates a structure member feature.

 Adds a gusset feature between 2 planar adjoining faces.

 Creates an end cap feature.

 Adds a fillet weld bead feature.

 Trims or extends structure members.

 Weld bead.

Quick Reference Guide to SOLIDWORKS Command Icons & Toolbars

DIMENSIONS/RELATIONS Toolbar

 Inserts dimension between two lines.

 Creates a horizontal dimension between selected entities.

 Creates a vertical dimension between selected entities.

 Creates a reference dimension between selected entities.

 Creates a set of ordinate dimensions.

 Creates a set of Horizontal ordinate

 Creates a set of Vertical ordinate dimensions.

 Creates a chamfer dimension.

 Adds a geometric relation.

 Automatically Adds Dimensions to the current sketch.

 Displays and deletes geometric relations.

 Fully defines a sketch.

 Scans a sketch for elements of equal length or radius.

 Angular Running dimension.

 Display / Delete dimension.

 Isolate changed dimension.

 Path length dimension.

BLOCK Toolbar

 Makes a new block.

 Edits the selected block.

 Inserts a new block to a sketch or drawing.

 Adds/Removes sketch entities to/from blocks.

 Updates parent sketches affected by this block.

 Saves the block to a file.

 Explodes the selected block.

 Inserts a belt.

STANDARD VIEWS Toolbar

 Front view.

 Back view.

 Left view.

 Right view.

 Top view.

 Bottom view.

 Isometric view.

 Trimetric view.

 Dimetric view.

 Normal to view.

 Links all views in the viewport together.

 Displays viewport with front & right

 Displays a 4 view viewport with 1st or 3rd

 Displays viewport with front & top.

 Displays viewport with a single view.

 View selector.

 New view.

FEATURES Toolbar

 Creates a boss feature by extruding a sketched profile.

 Creates a revolved feature based on profile and angle parameter.

 Creates a cut feature by extruding a sketched profile.

 Creates a cut feature by revolving a sketched profile.

 Thread.

 Creates a cut by sweeping a closed profile along an open or closed path.

 Loft cut.

 Creates a cut by thickening one or more adjacent surfaces.

 Adds a deformed surface by push or pull on points.

 Creates a lofted feature between two or more profiles.

 Creates a solid feature by thickening one or more adjacent surfaces.

 Creates a filled feature.

 Chamfers an edge or a chain of tangent edges.

 Inserts a rib feature.

 Combine.

 Creates a shell feature.

 Applies draft to a selected surface.

 Creates a cylindrical hole.

 Inserts a hole with a pre-defined cross section.

 Puts a dome surface on a face.

 Model break view.

 Applies global deformation to solid or surface bodies.

 Wraps closed sketch contour(s) onto a face.

 Curve Driven pattern.

 Suppresses the selected feature or component.

 Un-suppresses the selected feature or component.

 Flexes solid and surface bodies.

 Intersect.

 Variable Patterns.

 Live Section Plane.

 Mirrors.

 Scale.

 Creates a Sketch Driven pattern.

 Creates a Table Driven Pattern.

 Inserts a split Feature.

 Hole series.

 Joins bodies from one or more parts into a single part in the context of an assembly.

 Deletes a solid or a surface.

 Instant 3D.

 Inserts a part from file into the active part document.

 Moves/Copies solid and surface bodies or moves graphics bodies.

 Merges short edges on faces.

 Pushes solid / surface model by another solid / surface model.

 Moves face(s) of a solid.

 FeatureWorks Options.

 Linear Pattern.

 Fill Pattern.

 Cuts a solid model with a

 Boundary Boss/Base.

 Boundary Cut.

 Circular Pattern.

 Recognize Features.

 Grid System.

MOLD TOOLS Toolbar

 Extracts core(s) from existing tooling split.

 Constructs a surface patch.

 Moves face(s) of a solid.

 Creates offset surfaces.

 Inserts cavity into a base part.

 Scales a model by a specified factor.

 Applies draft to a selected surface.

 Inserts a split line feature.

 Creates parting lines to separate core & cavity surfaces.

 Finds & creates mold shut-off surfaces.

 Creates a planar surface from a sketch or a set of edges.

 Knits surfaces together.

 Inserts ruled surfaces from edges.

 Creates parting surfaces between core & cavity surfaces.

 Creates multiple bodies from a single body.

 Inserts a tooling split feature.

 Creates parting surfaces between the core & cavity.

 Inserts surface body folders for mold operation.

SELECTION FILTERS Toolbar

 Turns selection filters on and off.

 Clears all filters.

 Selects all filters.

 Inverts current selection.

 Allows selection of edges only.

 Allows selection filter for vertices only.

 Allows selection of faces only.

 Adds filter for Surface Bodies.

 Adds filter for Solid Bodies.

 Adds filter for Axes.

 Adds filter for Planes.

 Adds filter for Sketch Points.

 Allows selection for sketch only.

 Adds filter for Sketch Segments.

 Adds filter for Midpoints.

 Adds filter for Center Marks.

 Adds filter for Centerline.

 Adds filter for Dimensions and Hole Callouts.

 Adds filter for Surface Finish Symbols.

 Adds filter for Geometric Tolerances.

 Adds filter for Notes / Balloons.

 Adds filter for Weld Symbols.

 Adds filter for Weld beads.

 Adds filter for Datum Targets.

 Adds filter for Datum feature only.

 Adds filter for blocks.

 Adds filter for Cosmetic Threads.

Adds filter for Dowel pin symbols.

 Adds filter for connection points.

 Adds filter for routing points.

SOLIDWORKS Add-Ins Toolbar

 Loads/unloads CircuitWorks add-in.

 Loads/unloads the Design Checker add-in.

 Loads/unloads the PhotoView 360 add-in.

 Loads/unloads the Scan-to-3D add-in.

 Loads/unloads the SOLIDWORKS Motions add-in.

 Loads/unloads the SOLIDWORKS Routing add-in.

 Loads/unloads the SOLIDWORKS Simulation add-in.

 Loads/unloads the SOLIDWORKS Toolbox add-in.

 Loads/unloads the SOLIDWORKS TolAnalysis add-in.

 Loads/unloads the SOLIDWORKS Flow Simulation add-in.

 Loads/unloads the SOLIDWORKS Plastics add-in.

 Loads/unloads the SOLIDWORKS MBD SNL license.

FASTENING FEATURES Toolbar

 Creates a parameterized mounting boss.

 Creates a parameterized snap hook.

 Creates a groove to mate with a hook feature.

 Uses sketch elements to create a vent for air flow.

 Creates a lip/groove feature.

SCREEN CAPTURE Toolbar

 Copies the current graphics window to the clipboard.

 Records the current graphics window to an AVI file.

 Stops recording the current graphics window to an AVI file.

EXPLODE LINE SKETCH Toolbar

 Adds a route line that connect entities.

 Adds a jog to the route lines.

LINE FORMAT Toolbar

 Changes layer properties.

 Changes the current document layer.

 Changes line color.

 Changes line thickness.

 Changes line style.

 Hides / Shows a hidden edge.

 Changes line display mode.

Did you know??

* Ctrl+Q will force a rebuild on all features of a part.

* Ctrl+B will rebuild the feature being worked on and its dependents.

2D-To-3D Toolbar

 Makes a Front sketch from the selected entities.

 Makes a Top sketch from the selected entities.

 Makes a Right sketch from the selected entities.

- Makes a Left sketch from the selected entities.
- Makes a Bottom sketch from the selected entities.
- Makes a Back sketch from the selected entities.
- Makes an Auxiliary sketch from the selected entities.
- Creates a new sketch from the selected entities.
- Repairs the selected sketch.
- Aligns a sketch to the selected point.
- Creates an extrusion from the selected sketch segments, starting at the selected sketch point.
- Creates a cut from the selected sketch segments, optionally starting at the selected sketch point.

ALIGN Toolbar

- Aligns the left side of the selected annotations with the leftmost annotation.
- Aligns the right side of the selected annotations with the rightmost annotation.
- Aligns the top side of the selected annotations with the topmost annotation.
- Aligns the bottom side of the selected annotations with the lowermost annotation.
- Evenly spaces the selected annotations horizontally.
- Evenly spaces the selected annotations vertically.
- Centrally aligns the selected annotations horizontally.
- Centrally aligns the selected annotations vertically.
- Compacts the selected annotations horizontally.
- Compacts the selected annotations vertically.
- Creates a group from the selected items.
- Deletes the grouping between these items.
- Aligns & groups selected dimensions along a line or an arc.
- Aligns & groups dimensions at uniform distances.

- Evenly spaces selected dimensions.
- Aligns collinear selected dimensions.
- Aligns stagger selected dimensions.

SOLIDWORKS MBD Toolbar

- Captures 3D view.
- Manages 3D PDF templates.
- Creates shareable 3D PDF presentations.
- Toggles dynamic annotation views.

MACRO Toolbar

- Runs a Macro.
- Stops Macro recorder.
- Records (or pauses recording of) actions to create a Macro.
- Launches the Macro Editor and begins editing a new macro.
- Opens a Macro file for editing.
- Creates a custom macro.

SMARTMATES icons

- Concentric & Coincident 2 circular edges.
- Concentric 2 cylindrical faces.
- Coincident 2 linear edges.
- Coincident 2 planar faces.
- Coincident 2 vertices.
- Coincident 2 origins or coordinate systems.

TABLE Toolbar

 Adds a hole table of selected holes from a specified origin datum.

 Adds a Bill of Materials.

 Adds a revision table.

 Displays a Design table in a drawing.

 Adds a weldments cuts list table.

 Adds a Excel based of Bill of Materials

 Adds a weldment cut list table.

REFERENCE GEOMETRY Toolbar

 Adds a reference plane.

 Creates an axis.

 Creates a coordinate system.

 Adds the center of mass.

 Specifies entities to use as references using SmartMates.

SPLINE TOOLS Toolbar

 Inserts a point to a spline.

 Displays all points where the concavity of selected spline changes.

 Displays minimum radius of selected spline.

 Displays curvature combs of selected spline.

 Reduces numbers of points in a selected spline.

 Adds a tangency control.

 Adds a curvature control.

 Adds a spline based on selected sketch entities & edges.

 Displays the spline control polygon.

ANNOTATIONS Toolbar

 Inserts a note.

 Inserts a surface finish symbol.

 Inserts a new geometric tolerancing symbol.

 Attaches a balloon to the selected edge or face.

 Adds balloons for all components in selected view.

 Inserts a stacked balloon.

 Attaches a datum feature symbol to a selected edge / detail.

 Inserts a weld symbol on the selected edge / face / vertex.

 Inserts a datum target symbol and / or point attached to a selected edge / line.

 Selects and inserts block.

 Inserts annotations & reference geometry from the part / assembly into the selected.

 Adds center marks to circles on model.

 Inserts a Centerline.

 Inserts a hole callout.

 Adds a cosmetic thread to the selected cylindrical feature.

 Inserts a Multi-Jog leader.

 Selects a circular edge or an arc for Dowel pin symbol insertion.

 Adds a view location symbol.

 Inserts latest version symbol.

 Adds a cross hatch patterns or solid fill.

 Adds a weld bead caterpillar on an edge.

 Adds a weld symbol on a selected entity.

 Inserts a revision cloud.

 Inserts a magnetic line.

 Hides/shows annotation.

DRAWINGS Toolbar

 Updates the selected view to the model's current stage.

 Creates a detail view.

 Creates a section view.

 Inserts an Alternate Position view.

 Unfolds a new view from an existing view.

 Generates a standard 3-view drawing (1st or 3rd angle).

 Inserts an auxiliary view of an inclined surface.

 Adds an Orthogonal or Named view based on an existing part or assembly.

 Adds a Relative view by two orthogonal faces or planes.

 Adds a Predefined orthogonal projected or Named view with a model.

 Adds an empty view.

 Adds vertical break lines to selected view.

 Crops a view.

 Creates a Broken-out section.

QUICK SNAP Toolbar

 Snap to points.

 Snap to center points.

 Snap to midpoints.

 Snap to quadrant points.

 Snap to intersection of 2 curves.

 Snap to nearest curve.

 Snap tangent to curve.

 Snap perpendicular to curve.

Snap parallel to line.

 Snap horizontally / vertically to points.

 Snap horizontally / vertically.

 Snap to discrete line lengths.

 Snap to angle.

LAYOUT Toolbar

 Creates the assembly layout sketch.

 Sketches a line.

 Sketches a corner rectangle.

 Sketches a circle.

 Sketches a 3 point arc.

 Rounds a corner.

 Trims or extends a sketch.

 Adds sketch entities by offsetting faces, edges curves.

 Mirrors selected entities about a centerline.

 Adds a relation.

 Creates a dimension.

 Displays / Deletes geometric relations.

 Makes a new block.

 Edits the selected block.

 Inserts a new block to the sketch or drawing.

 Adds / Removes sketch entities to / from a block.

 Saves the block to a file.

 Explodes the selected block.

 Creates a new part from a layout sketch block.

 Positions 2 components relative to one another.

CURVES Toolbar

 Projects sketch onto selected surface.

 Inserts a split line feature.

 Creates a composite curve from selected edges, curves and sketches.

 Creates a curve through free points.

 Creates a 3D curve through reference points.

 Helical curve defined by a base sketch and shape parameters.

VIEW Toolbar

 Displays a view in the selected orientation.

 Reverts to previous view.

 Redraws the current window.

 Zooms out to see entire model.

 Zooms in by dragging a bounding box.

 Zooms in or out by dragging up or down.

 Zooms to fit all selected entities.

 Dynamic view rotation.

 Scrolls view by dragging.

 Displays image in wireframe mode.

 Displays hidden edges in gray.

 Displays image with hidden lines removed.

 Controls the visibility of planes.

 Controls the visibility of axis.

 Controls the visibility of parting lines.

 Controls the visibility of temporary axis.

 Controls the visibility of origins.

 Controls the visibility of coordinate systems.

 Controls the visibility of coordinate systems.

 Controls the visibility of reference curves.

 Controls the visibility of sketches.

 Controls the visibility of 3D sketch planes.

 Controls the visibility of 3D sketch.

 Controls the visibility of all annotations.

 Controls the visibility of reference points.

 Controls the visibility of routing points.

 Controls the visibility of lights.

 Controls the visibility of cameras.

 Controls the visibility of sketch relations.

 Changes the display state for the current configuration.

 Rolls the model view.

 Turns the orientation of the model view.

 Dynamically manipulate the model view in 3D to make selection.

 Changes the display style for the active view.

 Displays a shade view of the model with its edges.

 Displays a shade view of the model.

 Toggles between draft quality & high quality HLV.

 Cycles through or applies a specific scene.

 Views the models through one of the model's cameras.

 Displays a part or assembly w/different colors according to the local radius of curvature.

 Displays zebra stripes.

 Displays a model with hardware accelerated shades.

 Applies a cartoon affect to model edges & faces.

 Views simulations symbols.

TOOLS Toolbar

 Calculates the distance between selected items.

 Adds or edits equation.

 Calculates the mass properties of the model.

 Checks the model for geometry errors.

 Inserts or edits a Design Table.

 Evaluates section properties for faces and sketches that lie in parallel planes.

 Reports Statistics for this Part/Assembly.

 Deviation Analysis.

 Runs the SimulationXpress analysis wizard powered by SOLIDWORKS Simulation.

 Checks the spelling.

 Import diagnostics.

 Runs the DFMXpress analysis wizard.

 Runs the SOLIDWORKSFloXpress analysis wizard.

ASSEMBLY Toolbar

 Creates a new part & inserts it into the assembly.

 Adds an existing part or sub-assembly to the assembly.

 Creates a new assembly & inserts it into the assembly.

 Turns on/off large assembly mode for this document.

 Hides / shows model(s) associated with the selected model(s).

 Toggles the transparency of components.

 Changes the selected components to suppressed or resolved.

 Inserts a belt.

 Toggles between editing part and assembly.

 Smart Fasteners.

 Positions two components relative to one another.

 External references will not be created.

 Moves a component.

 Rotates an un-mated component around its center point.

 Replaces selected components.

 Replaces mate entities of mates of the selected components on the selected Mates group.

 Creates a New Exploded view.

 Creates or edits explode line sketch.

 Interference detection.

 Shows or Hides the Simulation toolbar.

 Patterns components in one or two linear directions.

 Patterns components around an axis.

 Sets the transparency of the components other than the one being edited.

 Sketch driven component pattern.

 Pattern driven component pattern.

 Curve driven component pattern.

 Chain driven component pattern.

 SmartMates by dragging & dropping components.

 Checks assembly hole alignments.

 Mirrors subassemblies and parts.

SOLIDWORKS Quick-Guide©
STANDARD Keyboard Shortcuts

Rotate the model

* Horizontally or Vertically: _____ Arrow keys

* Horizontally or Vertically 90°: _____ Shift + Arrow keys

* Clockwise or Counterclockwise: _____ Alt + left or right Arrow

* Pan the model: _____ Ctrl + Arrow keys

* Zoom in: _____ Z (shift + Z or capital Z)

* Zoom out: _____ z (lower case z)

* Zoom to fit: _____ F

* Previous view: _____ Ctrl+Shift+Z

View Orientation

* View Orientation Menu: _____ Space bar

* Front: _____ Ctrl+1

* Back: _____ Ctrl+2

* Left: _____ Ctrl+3

* Right: _____ Ctrl+4

* Top: _____ Ctrl+5

* Bottom: _____ Ctrl+6

* Isometric: _____ Ctrl+7

Selection Filter & Misc.

* Filter Edges: _____ e

* Filter Vertices: _____ v

* Filter Faces: _____ x

* Toggle Selection filter toolbar: _____ F5

* Toggle Selection Filter toolbar (on/off): _____ F6

* New SOLIDWORKS document: _____ F1

* Open Document: _____ Ctrl+O

* Open from Web folder: _____ Ctrl+W

* Save: _____ Ctrl+S

* Print: _____ Ctrl+P

* Magnifying Glass Zoom _____ g

* Switch between the SOLIDWORKS documents _____ Ctrl + Tab

SOLIDWORKS Sample Customized Hot Keys

Function Keys

Key	Command
F1	SW-Help
F2	2D Sketch
F3	3D Sketch
F4	Modify
F5	Selection Filters
F6	Move (2D Sketch)
F7	Rotate (2D Sketch)
F8	Measure
F9	Extrude
F10	Revolve
F11	Sweep
F12	Loft

Sketch

Key	Command
C	Circle
P	Polygon
E	Ellipse
O	Offset Entities
Alt + C	Convert Entities
M	Mirror
Alt + M	Dynamic Mirror
Alt + F	Sketch Fillet
T	Trim
Alt + X	Extend
D	Smart Dimension
Alt + R	Add Relation
Alt + P	Plane
Control + F	Fully Define Sketch
Control + Q	Exit Sketch